Target
Get back on track

GRADE

Edexcel GCSE (9–1)
Mathematics
Algebra

Diane Oliver

Pearson

Published by Pearson Education Limited, 80 Strand, London, WC2R ORL.

www.pearsonschoolsandfecolleges.co.uk

Text © Pearson Education Limited 2017
Typeset by Tech-Set Ltd, Gateshead
Original illustrations © Pearson Education Ltd 2017

The right of Diane Oliver to be identified as author of this work has been asserted by her in accordance with the Copyright, Designs and Patents Act 1988.

First published 2017

19 18 17
10 9 8 7 6 5 4 3 2 1

British Library Cataloguing in Publication Data
A catalogue record for this book is available from the British Library

ISBN 978 0 435 18337 0

Printed in Italy by Lego S.p.A

Helping you to formulate grade predictions, apply interventions and track progress.

Any reference to indicative grades in the Pearson Target Workbooks and Pearson Progression Services is not to be used as an accurate indicator of how a student will be awarded a grade for their GCSE exams.

You have told us that mapping the Steps from the Pearson Progression Maps to indicative grades will make it simpler for you to accumulate the evidence to formulate your own grade predictions, apply any interventions and track student progress.

We're really excited about this work and its potential for helping teachers and students. It is, however, important to understand that this mapping is for guidance only to support teachers' own predictions of progress and is not an accurate predictor of grades.

Our Pearson Progression Scale is criterion referenced. If a student can perform a task or demonstrate a skill, we say they are working at a certain Step according to the criteria. Teachers can mark assessments and issue results with reference to these criteria which do not depend on the wider cohort in any given year. For GCSE exams however, all Awarding Organisations set the grade boundaries with reference to the strength of the cohort in any given year. For more information about how this works please visit: https://qualifications.pearson.com/en/support/support-topics/results-certification/understanding-marks-and-grades.html/Teacher

Each practice question features a Step icon which denotes the level of challenge aligned to the Pearson Progression Map and Scale.

To find out more about the Progression Scale for Maths and to see how it relates to indicative GCSE 9–1 grades go to www.pearsonschools.co.uk/ProgressionServices

Contents

Useful formulae

Unit 8 Pre-calculus

Gradient of a straight line $= \dfrac{\text{change in } y}{\text{change in } x}$

Glossary

Unit 1 Graphs

Cubic function: a function in which the highest power of x is x^3. A cubic function can be written in the form $y = ax^3 + bx^2 + cx + d$.

Coefficient: the number in front of a variable.

Sketch (graph): depiction of a graph showing the general shape and the points of intersection of the graph with the axes.

Root (of an equation): a solution to the equation. The roots of a cubic equation are the x coordinates of the points where the curve crosses the x-axis.

Unit 2 Graphical inequalities

Inequality: a relationship between two values that are different, e.g. an expression is greater than a number or a variable is less than or equal to a number.

Solution set: the range of values that satisfy an inequality or all the solutions to an equation.

Integer: a positive or negative whole number, or zero.

Unit 3 Exponential graphs

Exponential function: a function that increases or decreases by the same multiplier. An exponential function can be written in the form $y = ab^x$ where $a > 0$.

Exponential graph: the graph of an exponential function. An exponential graph does not touch the x-axis. Exponential graphs of the form $y = a^x$ cross the y-axis at 1. Exponential graphs of the form $y = ab^x$ cross the y-axis at a.

Interest: money that is paid on a sum of money when it is borrowed or invested, usually expressed as a percentage.

Compound interest: when interest is paid on interest earned in the previous period in addition to the original amount.

Unit 4 Simultaneous equations

Simultaneous equations: equations in two (or more) variables that are true at the same time.

Linear equation: an equation in which the highest power of x is x. A linear equation can be written in the form $y = mx + c$.

Quadratic equation: an equation in which the highest power of x is x^2. A quadratic equation can be written in the form $y = ax^2 + bx + c$.

Solve: work out the value(s) of the variable(s) that make the equation(s) true.

Solution: the value(s) of the variable(s) that make the equation(s) true.

Graphical method: a method that uses the drawing of graphs to find an approximate solution to a problem. For example to solve simultaneous equations using a graphical method, you draw the graphs of the equations on the same axes and the solution is where the graphs intersect.

Unit 5 Trigonometric graphs

Trigonometric function: a function involving one of the trigonometric ratios of sides in a right angled triangle: sine, cosine or tangent, for example $y = \sin x$ $y = \cos x$ or $y = \tan x$.

Unit 6 Functions

Function: a rule using values of x to work out values of y. The function of x is written as f(x).

Inverse function: the function that 'undoes' the effect of a function. The inverse function of x is written as f^{-1}(x).

Composite function: a function in which one function is applied to the result of another. To work out a composite function, fg(x), substitute g(x) in f(x).

Unit 7 Transformations of graphs

Transformation: a way to change the position of a point, line or shape, or the dimensions of a shape.

Translation: a transformation that moves a point, line or shape a certain distance in a particular direction.

Reflection: a transformation that moves a point, line or shape to the opposite side of the line of reflection; the image of each point is the same distance from the line of reflection as the original point but on the opposite side.

Vertical stretch: a transformation in which the x-coordinates of the points remain the same and the y-coordinates are multiplied by a scale factor.

Horizontal stretch: a transformation in which the y-coordinates of the points remain the same and the x-coordinates are multiplied by a scale factor.

Scale factor: the number that you multiply by in a stretch or enlargement.

Unit 8 Pre-calculus

Gradient: the rate of change of y with respect to x; the steepness of the line or curve.

Distance–time graph: a graph with distance on the vertical axis and time on the horizontal axis; the gradient of the tangent to a point on a distance–time graph gives the speed at that point.

Speed–time graph: a graph with speed on the vertical axis and time on the horizontal axis; the gradient of the tangent to a point on a speed–time graph gives the acceleration at that point; the area under part of a speed–time graph gives the distance travelled in that time interval.

Speed: rate of change of distance with time.

Velocity: similar to speed but it is the rate of change in distance travelled in a particular direction with time.

Acceleration: rate of change of speed with time.

Non-linear: not a straight line.

Tangent: a line that just touches a curve.

① Graphs

This unit will help you to understand and use graphs of cubic functions.

① Match each equation to the correct graph.

a $y = x^2$ ....

b $y = -x^2$ ....

c $y = (x + 2)(x - 5)$ ....

d $y = (2x - 1)(x + 5)$ ....

e $y = (x - 2)(x + 5)$ ....

f $y = (x + 2)(x + 5)$ ....

i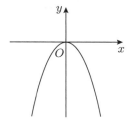

ii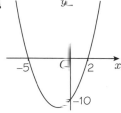

iii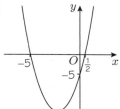

iv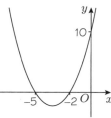

v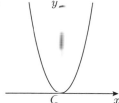

vi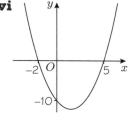

Key points

The highest power of x in a cubic function is x^3.

A cubic function can be written in the form $y = ax^3 + bx^2 + cx + d$

These **skills boosts** will help you to recognise, interpret and sketch graphs of cubic functions.

1 Recognising and interpreting graphs of cubic functions

2 Sketching cubic graphs

You might have already done some work on cubic graphs. Before starting the first skills boost, rate your confidence using each concept.

① Draw the graph of $y = x^3 - 3$ and use it to solve the equation $x^3 - 3 = 2$

② Sketch the graph of $y = (x + 3)(x - 2)(x + 1)$

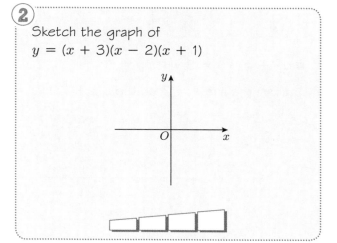

How confident are you?

1 Recognising and interpreting graphs of cubic functions

Graphs of cubic functions, $y = ax^3 + bx^2 + cx + d$, when a is positive, look like ⟋ or ⟋ ⟍

Graphs of cubic functions, $y = ax^3 + bx^2 + cx + d$, when a is negative, look like ⟍ or ⟍ ⟋

Guided practice

 Worked exam question

a Complete this table of values for $y = x^3 + x - 1$

x	-2	-1	0	1	2
y		-3			9

b Draw the graph of $y = x^3 + x - 1$

c Use your graph to solve the equation $x^3 + x - 1 = 3$

---

a Substitute each value of x in the equation.

When $x = -2$, $y = (-2)^3 + (-2) - 1 = $ ....

When $x = 0$, $y = (0)^3 + (0) - 1 = $ ....

When $x = 1$, $y = (1)^3 + (1) - 1 = $ ....

Write the values in the table.

x	-2	-1	0	1	2
y	-11	-3	-1	1	9

b Use the coordinates in the table to plot the points for the graph.
Join your points with a smooth curve.

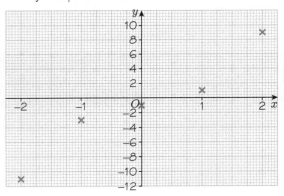

c $x^3 + x - 1 = 3$ when $y = 3$

Draw the line $y = 3$ on the graph.

Where the line and the curve intersect is the solution to the equation $x^3 + x - 1 = 3$

$x = 1.4$

A range of acceptable values for x is
$1.35 \leqslant x \leqslant 1.45$

1 Circle the graphs of cubic functions.

A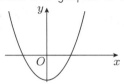

B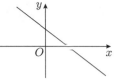

C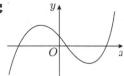

D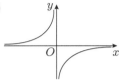

E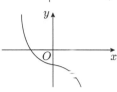

F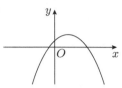

2 a Complete this table of values for $y = x^3 - 5x + 3$

x	−2	−1	0	1	2	3
y	....	7	....	....	1	....

b Draw the graph of $y = x^3 - 5x + 3$

c Use your graph to solve the equation $x^3 - 5x + 3 = 5$

....

3 a Complete this table of values for $y = 2x^3 - 7x - 1$

x	−2	−1	0	1	2
y	−3	....	....	−6	....

b Draw the graph of $y = 2x^3 - 7x - 1$

c Use your graph to solve the equation $2x^3 - 7x - 1 = 0$

....

Exam-style question

4 a Complete this table of values for $y = x^3 - 4x - 3$

x	−2	−1	0	1	2	3
y	....	0	....	....	−3	....

(2 marks)

b Draw the graph of $y = x^3 - 4x - 3$　(2 marks)

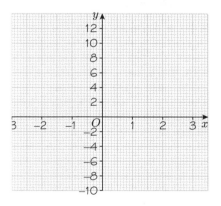

c Use your graph to solve the equation $x^3 - 4x - 3 = 0$

.... (2 marks)

Reflect　Why does the coefficient of x^3 determine whether the cubic graph starts by increasing or decreasing?

② Sketching cubic graphs

Guided practice

Sketch the graph of

a $y = (x - 3)(x - 1)(x + 2)$ **b** $y = -(x - 2)^2(x + 1)$

---

a Let $y = 0$ to find where the graph crosses the x-axis, the roots of the equation.

When $y = 0$, $0 = (x - 3)(x - 1)(x + 2)$

$(x - 3) = 0$, so $x = 3$ or .... $= 0$, so $x = $ .... or .... $= 0$, so $x = $ ....

The roots of the equation are $x = 3$, $x = $ .... and $x = $ ....

The curve crosses the x-axis at the points $(3, 0)$, $(............, ...)$ and $(............, ...)$.

> You find the roots by solving the equation when $y = 0$
> Put each bracket equal to zero.

Let $x = 0$ to find where the graph crosses the y-axis.

$y = (0 - 3)(0 - 1)(0 + 2) = 6$

> The graph crosses the y-axis when $x = 0$

The curve crosses the y-axis at $(............, ...)$.

$y = (x - 3)(x - 1)(x + 2)$
$\quad = (x - 3)(x^2 + x - 2)$

$\quad = x^3$ ....

> The coefficient of x^3 is 1, which is positive, so the shape of the graph is

Sketch the graph.

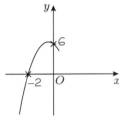

> Label where the graph crosses the x- and y-axes.

b Work out where the graph crosses the x-axis by finding the roots of the equation.

When $y = 0$, $0 = -(x - 2)^2(x + 1)$

$(x - 2) = 0$, so $x = 2$

or .... $= 0$, so $x = $ ....

> $x = 2$ is a repeated solution so this is where the curve touches the x-axis.
> Put each bracket equal to zero.

The roots of the equation are $x = 2$ and $x = $ ....

The curve touches the x-axis at $(2, 0)$ and crosses the x-axis at $(............, ...)$.

Let $x = 0$ to find where the graph crosses the y-axis.

$y = -(0 - 2)^2(0 + 1) = $ ....

So the curve crosses the y-axis at $(............, ...)$.

$y = -(x - 2)^2(x + 1)$
$\quad = -(x + 1)(x^2 - 4x + 4)$
$\quad = -x^3$ ....

> The coefficient of x^3 is 1, which is negative, so the shape of the graph is

Sketch the graph.

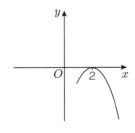

> Label where the graph intersects the x- and y-axes.

1. Match each equation to the correct graph.
   a $y = (x - 1)(x + 2)(x + 3)$ ....
   b $y = -(x - 1)(x + 2)(x + 3)$ ....
   c $y = (x - 1)(x - 2)(x - 3)$ ....
   d $y = x(x + 2)(x + 3)$ ....
   e $y = (x + 1)(x + 2)^2$ ....
   f $y = x^2(x + 3)$ ....

   i
   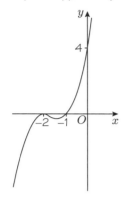

   ii
   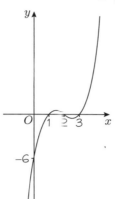

   iii
   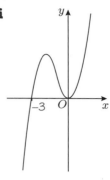

   iv
   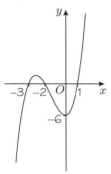

   v
   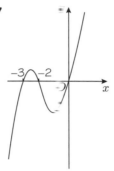

   vi
   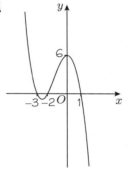

2. Sketch the graphs, marking clearly the points of intersection with the x- and y-axes.
   a $y = (x - 1)(x + 1)(x + 4)$
   b $y = (x - 2)(x - 1)(x + 2)$
   c $y = -x(x - 2)(x + 2)$

   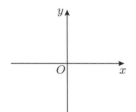
   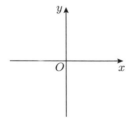
   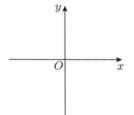

   d $y = (x - 1)(x + 1)(x + 4)$
   e $y = (x - 2)^2(x + 1)$
   f $y = -(x - 1)(x + 1)^2$

   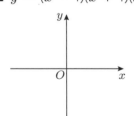
   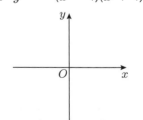
   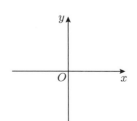

3. Sketch the graph of $y = -x(x - 3)^2$
   Clearly mark the points of intersection with the axes.

   (3 marks)

Reflect Explain how you know when the curve touches the x-axis and when it crosses the x-axis.

Practise the methods

Answer this question to check where to start.

Check up

Match each graph to its equation.

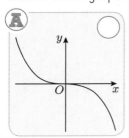 **A** ◯

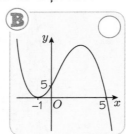 **B** ◯

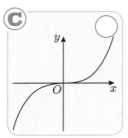 **C** ◯

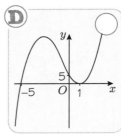 **D** ◯

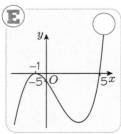 **E** ◯

1 $y = (x - 1)^2(x + 5)$

4 $y = -(x - 5)(x + 1)^2$

2 $y = x^3$

5 $y = -x^3$

3 $y = (x - 5)(x + 1)^2$

If you matched A5, B4, C2, D1 and E3 go on to Q3.

If you matched them differently go to Q1 for more practice.

① **a** **i** Write the roots of the equation $y = -(x + 1)(x - 1)(x + 5)$ ....

　　ii Write the coordinates where the graph crosses the x-axis. ....

　b Where does the graph of $y = -(x + 1)(x - 1)(x + 5)$ cross the y-axis? ....

　c Sketch the graph of $y = -(x + 1)(x - 1)(x + 5)$

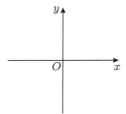

② Sketch each graph, marking the points of intersection with both axes.

　a $y = (x + 1)(x - 1)(x - 2)$

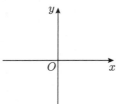

　b $y = (x + 1)(x - 2)^2$

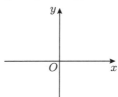

Exam-style question

③ **a** Complete this table of values for $y = x^3 - 7x + 2$

x	−3	−2	−1	0	1	2	3
y	....	....	8	....	....	−4	....

(2 marks)

　b Draw the graph of $y = x^3 - 7x + 2$ (on the grid opposite).

(2 marks)

　c Use your graph to solve the equation $x^3 - 7x + 2 = 0$

.... (2 marks)

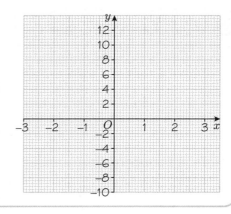

Problem-solve!

(1) This graph has equation $y = x^3 + ax^2 + bx + c$

Work out the values of a, b and c.

....

....

(2) A graph has equation $y = -x^3 + ax^2 + bx + c$

It crosses the x-axis at $x = 2$, $x = 5$ and $x = -1$

Without drawing the graph, work out the values of a, b and c.

....

Exam-style questions

(3) Here are four graphs.

A 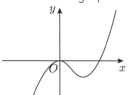 **B** 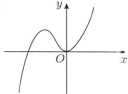 **C** 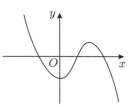 **D**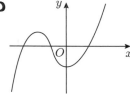

Here are five equations of graphs.

  i $y = -(x + 3)(x - 3)(x - 7)$ **ii** $y = -(x + 3)(x - 3)(x + 7)$ **iii** $y = x^2(x + 4)$

  iv $y = x^2(x - 4)$ **v** $y = (x + 5)(x + 2)(x - 4)$

a Match each graph to the correct equation.

.... (4 marks)

b There is one equation that you have not used.
  Sketch a graph for this equation.

(2 marks)

(4) This is the graph of $y = x^3 + 2x^2 - 3$

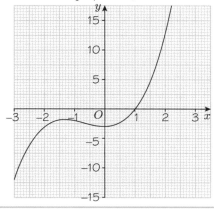

Use the graph to solve the equation $x^3 + 2x^2 - x = 0$

.... (3 marks)

Now that you have completed this unit, how confident do you feel?

1 Recognising and interpreting graphs of cubic functions

2 Sketching cubic graphs

② Graphical inequalities

This unit will help you to sketch and solve inequalities.

AO1 Fluency check

① Write the integer values of n that satisfy each inequality.

a $n \leqslant 7$ **b** $n > -4$ **c** $2n - 5 \geqslant -7$ **d** $19 < 2n + 3$

....

② Sketch the graph of

a $y = 3$ **b** $y = x + 2$ **c** $y = 2x - 1$ **d** $x + y = 4$

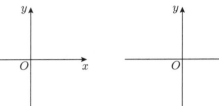
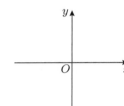
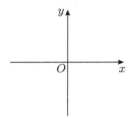

Key points

| The points that satisfy an inequality can be represented on a graph. | A dotted line on a graph means that the points on the line are not included. |

These **skills boosts** will help you to sketch linear inequalities and solve quadratic inequalities using graphs.

| **1** Linear inequalities in one variable | **2** Linear inequalities in two variables | **3** Quadratic inequalities |

You might have already done some work on solving inequalities. Before starting the first skills boost, rate your confidence using each concept.

①
On the grid, shade the region that satisfies the inequalities
$x \leqslant 3$ and $y > -1$

②
On the grid, shade the region that satisfies the inequalities
$y \geqslant x - 2$, $y < x$ and $y > 1$

③
Solve the inequality
$2x^2 + x - 15 > 0$

How confident are you?

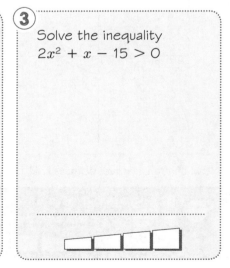

1 Linear inequalities in one variable

This shaded region satisfies the inequality $y > 2$.

This shaded region satisfies the inequality $x \leqslant 3$.

Guided practice

Shade the region that satisfies the inequalities $x > -2$ and $y \leqslant 4$

_____

Draw the dotted line $x = -2$

Draw the solid line $y = 4$

Shade the region to the right of $x > -2$ and below $y \leqslant 4$

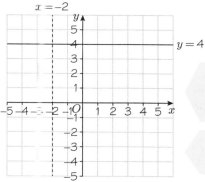

Worked exam question

All the points to the right of $x = -2$ have an x-value greater than -2.

All the points below $y = 4$ have a y-value less than 4.

① Write down the inequalities represented by each shaded region.

a

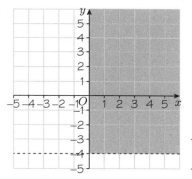

....

....

b

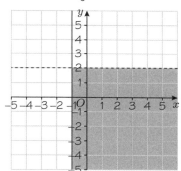

....

....

Exam-style question

② On the grid, shade the region that satisfies the inequalities

$x \geqslant -4$ and $y > 1$

Mark this region with the letter R.

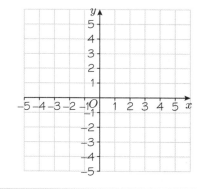

(2 marks)

Reflect How do you know whether to draw a dotted line or a solid line? How do you know where to draw the line? How do you know which side of the line to shade?

2 Linear inequalities in two variables

Guided practice

Shade the region that satisfies the inequalities

$x + y < 3 \qquad y \geqslant x \qquad x > -2$

Draw the dotted line $x + y = 3$

Choose a point on one side of the line to decide which side of the line to shade.

> At point (1, 1),
> $x + y = 1 + 1 =$ .... < 3,
> so the inequality is correct.

As the inequality is correct, you have identified the required region as being on the same side of the line as point (1, 1).

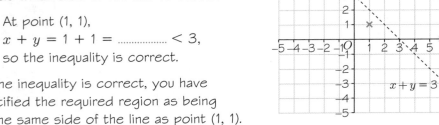

Do not shade the region until you have drawn all the lines.

Draw the solid line $y = x$

Choose a point on one side of the line to decide which side of the line is the required region.

> At point (1, 0),
>
> $y =$ .... $< x$, so the inequality is incorrect.

Draw the dotted line $x = -2$

Shade the region where the separate regions for each inequality overlap.

Choose a point in your shaded region and check that its coordinates satisfy all three inequalities.

① Write down the inequality represented by each shaded region.

a

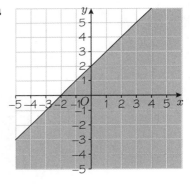

....

b

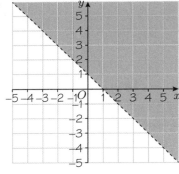

....

Hint

Choose a point on one side of the line to decide which inequality should be used.

② Write down the inequalities represented by each shaded region.

a

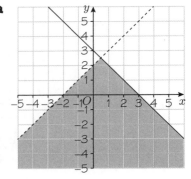

....
....

b

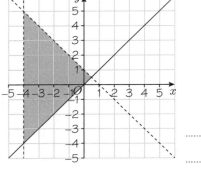

....
....

(3) Shade the region that satisfies the inequalities

a $x + y \leqslant 4$ and $y > -1$

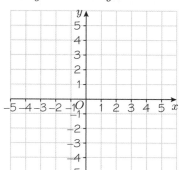

b $y > x + 3$ and $y \leqslant 2 - x$

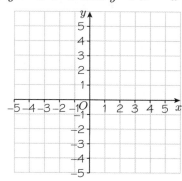

(4) Shade the region that satisfies the inequalities

a $y < 2$, $x + y > -2$ and $x \leqslant 2$

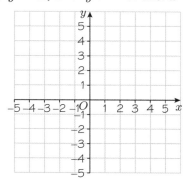

b $y < x$, $y \leqslant \frac{1}{2}x - 1$ and $y > 1 - x$

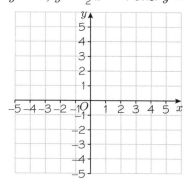

(5) Shade the region that satisfies the inequalities

a $x \geqslant 0$, $y < 3$, $x + y \leqslant 5$ and $y > -1$

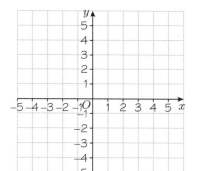

b $y \leqslant 2x + 3$, $x + y > -4$, $y > x - 4$ and $x + y \leqslant 4$

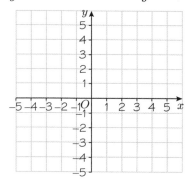

Exam-style question

(6) On the grid shown, shade the region that satisfies the inequalities

$x + y \leqslant 6$ $y < 2x$ $y > 2$

Label the region R.

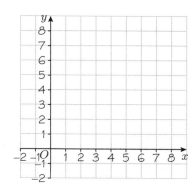

(4 marks)

Reflect   Is there a set of inequalities that does not have a region where they all overlap?

3 Quadratic inequalities

Guided practice

a Sketch the graph of $y = 2x^2 + 6x - 8$

b From the graph, identify the values of x for which $2x^2 + 6x - 8 \geqslant 0$

---

a Work out where the graph crosses the x-axis.

$$y = 2x^2 + 6x - 8$$
$$= 2(x^2 + 3x - 4)$$
$$= 2(x + 4)(x - 1)$$

When $y = 0$, $0 = 2(x + 4)(x - 1)$

So either $(x + 4) = 0$, giving $x = \dotsb$

or $(x - 1) = 0$, giving $x = \dotsb$

Work out where the graph crosses the y-axis.

When $x = 0$, $y = 2(0)^2 + 6(0) - 8 = \dotsb$

b Identify the x-values for which the graph is above the x-axis or on it.

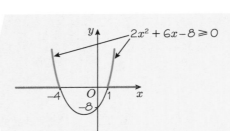

$x \leqslant \dotsb$ or $x \geqslant \dotsb$

If the quadratic inequality is $\leqslant$ or $\geqslant$, then the solutions will also use $\leqslant$ and $\geqslant$.

---

① **a** Sketch the graph of $y = x^2 - x - 6$

  b From the graph, identify the values of x for which $x^2 - x - 6 < 0$

  ....

② **a** Sketch the graph of $y = x^2 + 2x - 8$

  b Use the graph to solve $x^2 + 2x - 8 > 0$

  ....

③ **a** Sketch the graph of $y = x^2 + 7x + 12$

  b Hence solve the inequality $x^2 + 7x + 12 \geqslant 0$

  ....

④ Solve the inequality $x^2 + 5x < 6$

....

Hint
Before sketching the graph, rearrange the inequality to be < 0.

⑤ Solve

a $x^2 - 6x + 5 > 0$

....

b $2x^2 + 7x - 4 \leqslant 0$

....

c $-x^2 - 3x + 10 < 0$

....

d $2x^2 - 18 \geqslant 0$

....

e $x^2 + 2x - 15 \geqslant 0$

....

f $x^2 - 7x + 12 < 0$

....

Exam-style question

⑥ Solve $x^2 > 5x - 6$

.... (3 marks)

Reflect Explain why you cannot solve $(x - 1)(x - 2) < 0$ by solving $x - 1 < 0$ and $x - 2 < 0$

Practise the methods

Answer this question to check where to start.

Check up

Tick the region that satisfies the inequalities $y > 2$, $x + y \geq -1$ and $y < x - 2$

A ◯
B ◯
C ◯
D ◯

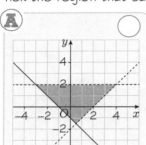 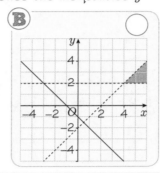 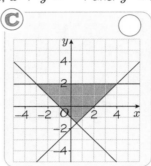 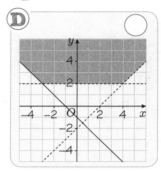

If you ticked B go to Q3.

If you ticked A, C or D go to Q1 for more practice.

① Write down the inequality represented by each shaded region.

a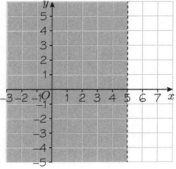

b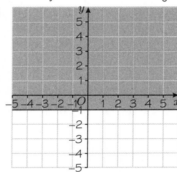

c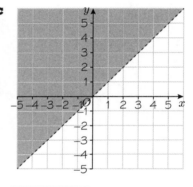

....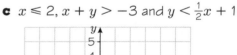

② On the grid, shade the region that satisfies the inequalities
 a $y < -2x$
 b $x + y \leq 0$ and $x > 1$
 c $x \leq 2$, $x + y > -3$ and $y < \frac{1}{2}x + 1$

 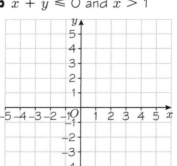 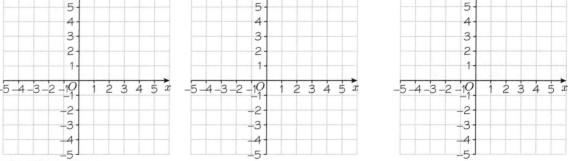

Exam-style question

③ Solve the inequality $x^2 \geq 4(x + 3)$

.... **(3 marks)**

Problem-solve!

① Work out how many points with integer coordinates satisfy the inequalities
$y > 3x - 4$, $y > -2x$ and $y < 3$

....

② Solve the inequality $3x^2 < 2(4 - 5x)$

.... (4 marks)

③ Plot the graph of $y = x^2 - 4x + 1$, to find, graphically, the range of values of x which satisfy the inequality $x^2 - 5x + 4 \leqslant 0$

....

Hint
Rearrange the inequality so the LHS is the same as the equation of the graph.

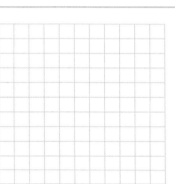

④ Plot the graph of $y = 2x^2 - x - 6$, to find, graphically, the range of values of x which satisfy the inequality $2x^2 - 3x - 5 > 0$

....

⑤ The area of the rectangle is less than 105 cm².

Work out the range of possible values for the length and width of the rectangle.

....

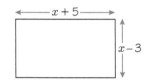

Now that you have completed this unit, how confident do you feel?

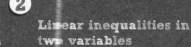
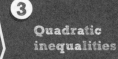

1 Linear inequalities in one variable

2 Linear inequalities in two variables

3 Quadratic inequalities

③ Exponential graphs

This unit will help you to recognise, sketch, draw and interpret exponential graphs.

AO1 Fluency check

① Match each equation to the correct graph.

a $y = x^3$ ....

b $y = -x^2$ ....

c $y = (x + 3)(x - 1)$ ....

d $y = (2x + 1)(x + 3)(x - 1)$ ....

e $y = (x - 1)(x + 3)^2$ ....

f $y = (x + 1)(x - 3)^2$ ....

i
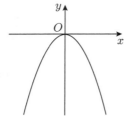

ii
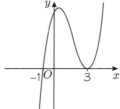

iii
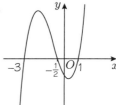

iv
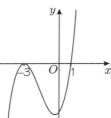

v
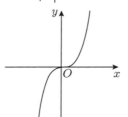

vi
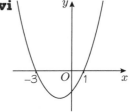

Key points

An exponential function is a function that increases or decreases by the same multiplier. For example, it keeps doubling.

An exponential function can be written in the form $y = a^x$, where $a > 0$

These **skills boosts** will help you to draw, use and recognise exponential graphs.

1 Drawing and interpreting exponential graphs of the form $y = a^x$

2 Recognising and sketching exponential graphs

3 Drawing and interpreting exponential graphs of the form $y = ab^x$

You might have already done some work on exponential graphs. Before starting the first skills boost, rate your confidence using each concept.

① Draw the graph of $y = 3^x$ and use it to solve the equation $3^x - 2 = 0$

② Sketch the graph of $y = 4^x$

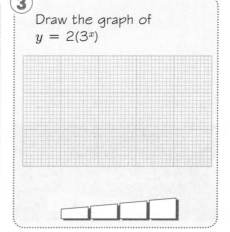

③ Draw the graph of $y = 2(3^x)$

How confident are you?

1 Drawing and interpreting exponential graphs of the form $y = a^x$

Guided practice

a Complete the table of values for $y = 2^x$

x	-3	-2	-1	0	1	2	3
y	$\frac{1}{8}$		$\frac{1}{2}$			4	

b Draw the graph of $y = 2^x$

c Use your graph to solve the equation $2^x = 3$

---

a Substitute each value of x in the equation.

When $x = -2$, $y = 2^{-2} = \dfrac{1}{2^2} =$ ....

When $x = 0$, $y = 2^0 =$ ....

When $x = 1$, $y = 2^1 =$ ....

When $x = 3$, $y = 2^3 =$ ....

Write the values in the table.

x	-3	-2	-1	0	1	2	3
y	$\frac{1}{8}$	$\frac{1}{4}$	$\frac{1}{2}$	1	2	4	8

b Use the coordinates in the table to plot the points for the graph.
Join your points with a smooth curve.

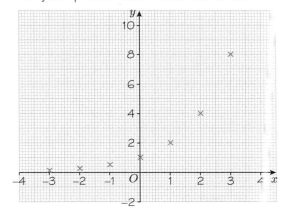

c $2^x = 3$ when $y = 3$

Draw the line $y = 3$ on your graph.

Where the line and the curve intersect is the solution to the equation $2^x = 3$

$x = 1.6$

A range of acceptable values for x is
$1.55 \leqslant x \leqslant 1.65$

① **a** Complete the table of values for $y = 3^x$

x	-2	-1	0	1	2	3
y	$\frac{1}{9}$	....	....	....	9	....

b Draw the graph of $y = 3^x$

c Use the graph to solve the equation $3^x = 5$ ....

② **a** Complete the table of values for $y = \left(\frac{2}{3}\right)^x$

x	-3	-2	-1	0	1	2	3
y	....	$\frac{9}{4}$	....	....	....	....	$\frac{8}{27}$

b Draw the graph of $y = \left(\frac{2}{3}\right)^x$

c Use the graph to solve the equation $\left(\frac{2}{3}\right)^x = 2$ ....

d Use the graph to solve the equation $y = \left(\frac{2}{3}\right)^{1.5}$ ....

③ **a** Complete this table of values for $y = 5^{-x}$

x	-2	-1	0	1	2
y	25	....	....	$\frac{1}{5}$	....

(2 marks)

b Draw the graph of $y = 5^{-x}$ (on the grid opposite).   (2 marks)

c Use your graph to solve the equation $5^{-x} = 10$

.... (1 mark)

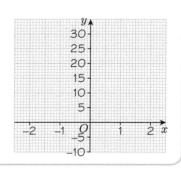

Reflect   Explain why all of your graphs on this page cross the y-axis at 1.

2 Recognising and sketching exponential graphs

The graph of an exponential function $y = a^x$, where $a > 1$, or $y = b^{-x}$, where $0 < b < 1$, looks like

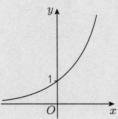

The graph of an exponential function $y = a^{-x}$, where $a > 1$, or $y = b^x$, where $0 < b < 1$, looks like

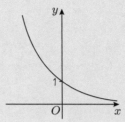

Guided practice

Sketch the graph of $y = 2^x$

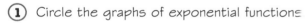

Identify which shape of graph you need.

Look at what happens to y for large negative and positive values of x.

When x is large and negative, 2^x is very small.

When x is large and positive, 2^x is very ....

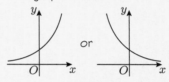

The graph will be

or

All graphs of the form $y = a^x$ cross the y-axis at (0, 1), since $a^0 = 1$
The graph crosses the x-axis when $y = 0$, but $2^x \neq 0$ so the graph doesn't cross the x-axis.

(1) Circle the graphs of exponential functions

A

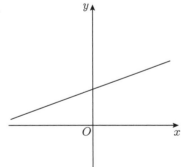

B

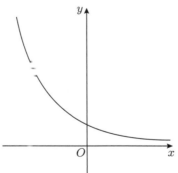

C

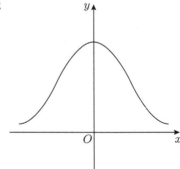

D

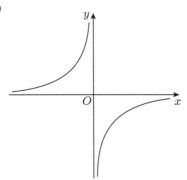

E

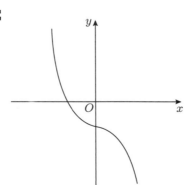

F

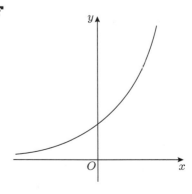

② Sketch the graph of $y = 4^x$, marking clearly the points of intersection with the axes.

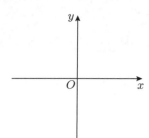

③ Sketch the graph of $y = 3^{-x}$, marking clearly the points of intersection with the axes.

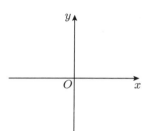

④ Sketch the graph of $y = \left(\frac{1}{2}\right)^x$, mark clearly the points of intersection with the axes.

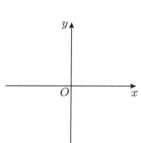

⑤ Sketch the graph of $y = \left(\frac{3}{4}\right)^{-x}$, marking clearly the points of intersection with the axes.

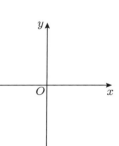

Exam-style question

⑥ Here are two graphs.

A 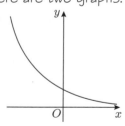   **B**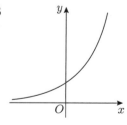

Here are four equations of graphs.

i $y = x^3$ **ii** $y = 10^x$ **iii** $y = 10^{-x}$ **iv** $y = \frac{1}{x}$

Match each graph to the correct equation.

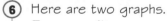

(2 marks)

Reflect   Compare the graphs of $y = 4^x$ and $\left(\frac{1}{4}\right)^{-x}$

Drawing and interpreting exponential graphs of the form $y = ab^x$

Guided practice

a Complete this table of values for $y = 3(2^{-x})$

x	−3	−2	−1	0	1	2	3
y	24	....	6	....	....	....	$\dfrac{3}{8}$

b Draw the graph of $y = 3(2^{-x})$

c Use your graph to solve the equation $3(2^{-x}) = 10$

---

a Substitute each value of x in the equation.

When $x = -2$, $y = 3(2^2) = $ ....

When $x = 0$, $y = 3(2^0) = $ ....

When $x = 1$, $y = 3(2^{-1}) = $ ....

When $x = 2$, $y = 3(2^{-2}) = \dfrac{3}{2^2} = $ ....

Write the values in the table.

x	−3	−2	−1	0	1	2	3
y	24	12	6	3	$\dfrac{3}{2}$	$\dfrac{3}{4}$	$\dfrac{3}{8}$

b Use the coordinates in the table to plot the points for the graph. Join your points with a smooth curve.

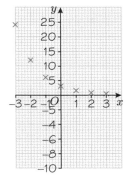

c $3(2^{-x}) = 10$ when $y = 10$

Draw the line $y = 10$ on your graph.

$x = -1.7$

> Where the line and the curve intersect is the solution to the equation $3(2^{-x}) = 10$

> A range of acceptable values for x is $1.70 \leqslant x \leqslant 1.80$

Exam-style question

1 **a** Complete this table of values for $y = 4\left(\dfrac{1}{3}\right)^x$

x	−2	−1	0	1	2
y	....	12	....	....	$\dfrac{4}{9}$

(2 marks)

b Draw the graph of $y = 4\left(\dfrac{1}{3}\right)^x$

(2 marks)

c Use your graph to solve the equation $4\left(\dfrac{1}{3}\right)^x = 3$

.... (1 mark)

Reflect

Before this page, graphs of exponential functions have crossed the y-axis at 1. Explain why the graphs of exponential functions on this page do not cross at 1. Where does the graph of $y = ab^x$ cross the y-axis?

Practise the methods

Answer this question to check where to start.

Check up

Cross out any graphs that are *not* exponential functions.

A ⃝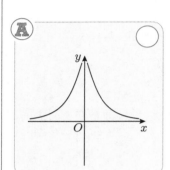

B ⃝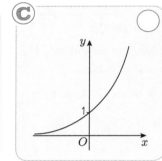

C ⃝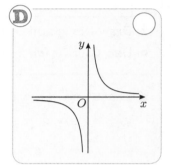

D ⃝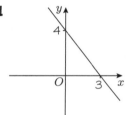

If you crossed out A and D, write possible functions for graphs B and C. Then go to Q2.

If you didn't cross out A and D go to Q1 for more practice.

1 Name the type of function shown in each graph.

a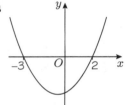

b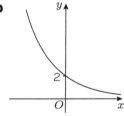

c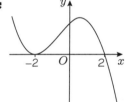

d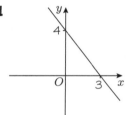

....

Exam-style question

2 **a** Complete this table of values for $y = \left(\frac{1}{2}\right)^{-x}$

x	-2	-1	0	1	2	3
y	....	$\frac{1}{2}$	....	....	4	....

(2 marks)

b Draw the graph of $y = \left(\frac{1}{2}\right)^{-x}$ (2 marks)

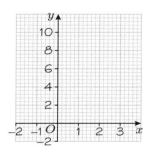

c Use your graph to solve the equation $\left(\frac{1}{2}\right)^{-x} = 5$

.... (1 mark)

Problem-solve!

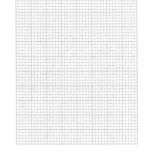

(1) **a** On the same axes, draw and label the graphs of

   i $y = 2.5^x$

   ii $y = 3.5^x$

   b Predict where the graph of $y = 3^x$ will be
   Sketch it on the same axes.

   c At which point do all the graphs intersect the y-axis?

   ....

(2) Vikki invests £500 at 2% compound interest per annum.

   a Write a formula for the value of the investment,
   I, and the number of years, t. .. ..

   b Draw a graph of I against t for the first 10 years.

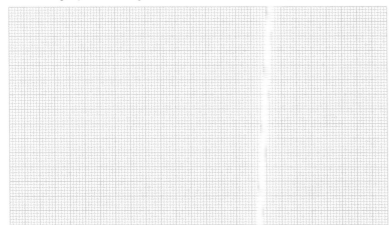

   c Use the graph to estimate when the investment will reach a value of £600. ....

Exam-style question

(3) The sketch shows a curve with equation $y = ka^x$,
   where k and a are constants and $a > 1$

   The curve passes through the points $(1, 4)$
   and $(3, 100)$.

   Calculate the value of k and the value of a.

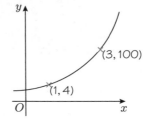

   .... **(3 marks)**

Now that you have completed this unit, how confident do you feel?

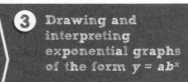

1 Drawing and interpreting exponential graphs of the form $y = a^x$

2 Recognising and sketching exponential graphs

3 Drawing and interpreting exponential graphs of the form $y = ab^x$

Simultaneous equations

This unit will help you to solve two equations simultaneously, where one equation is linear and the other non-linear.

AO1 Fluency check

(1) Make y the subject of each equation.

    a $y + 2x - 1 = 0$ **b** $2y - 4x = 10$ **c** $2x + 4y = 1$

(2) Expand and simplify

    a $(x + 5)(x - 3)$ **b** $(2x + 3)(x - 4)$ **c** $(3x + 1)(2x - 5)$

(3) Solve

    a $(3x + 2)(x - 4) = 0$ **b** $x^2 - 5x + 6 = 0$ **c** $2x^2 + 11x + 12 = 0$

Key points

Solving equations simultaneously means finding the values of all of the letters.	A linear equation can be written in the form $y = mx + c$	A quadratic equation can be written in the form $y = ax^2 + bx + c$	The equation of a circle can be written in the form $x^2 + y^2 = r^2$

These **skills boosts** will help you to solve two equations simultaneously, where one equation is linear and the other equation is either a quadratic or a circle.

> **1** Solving simultaneous equations graphically
>
> **2** Solving linear and non-linear simultaneous equations algebraically

You might have already done some work on solving simultaneous equations. Before starting the first skills boost, rate your confidence using each concept.

1

Use a graphical method to find an approximate solution to the simultaneous equations
$y = x^2 - 2x - 3$ and $x + y = 2$

2

Solve the simultaneous equations
a $y = 5x^2 + 5x + 1$ and $3x + y = 5$
b $x^2 + y^2 = 20$ and $2x + y = 3$
Give your answers to 2 decimal places (d.p.) where necessary.

How confident are you? 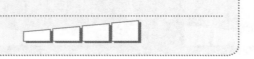   

1 Solving simultaneous equations graphically

Worked exam question

Guided practice

Use a graphical method to find approximate solutions to the simultaneous equations
$y + x^2 = 5$ and $y - 1 = x$

Rearrange the equations in the form '$y = ...$'.

$y + x^2 = 5$ $\qquad\qquad$ $y - 1 = x$

$\quad y =$ .... $\qquad$ $y =$ ....

Plot the graphs.

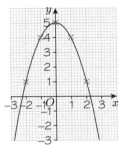

From the graph, read the coordinates of the points where the straight line intersects the quadratic.

The solutions are $x = -2.6$, $y = -1.6$

and $x =$ ...., $y =$ ....

The points of intersection are the solutions to the simultaneous equations, as these are the points that lie on both the curve and the line.

(1) Solve the simultaneous equations graphically.

$y = x^2 - 4$

$2y - x = 2$

....

....

Exam-style question

(2) **a** On the grid, construct the graph of $x^2 + y^2 = 25$

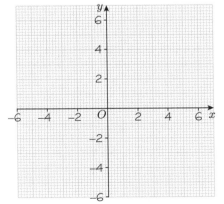

(2 marks)

b Find estimates for the solutions of the simultaneous equations

$x^2 + y^2 = 25$ $\qquad\qquad$ $y = 2x - 3$

.... (3 marks)

Reflect How can you draw your graphs to ensure your solutions are as accurate as possible?

2 Solving linear and non-linear simultaneous equations algebraically

A pair of linear and quadratic or circle simultaneous equations have two possible solutions providing the straight line intersects the curve or circle.

Guided practice

Solve these simultaneous equations.

a $y = 2x + 3$ and $y = 2x^2 - 3x - 9$

b $x^2 + y^2 = 13$ and $5x - y + 13 = 0$

---

a Substitute the linear expression for y into the quadratic equation.

$$2x + 3 = 2x^2 - 3x - 9$$

Rearrange so the right-hand side is 0.

$$2x^2 - 5x - 12 = 0$$

Factorise and solve the equation.

$(2x + 3)(\dots\dots\dots) = 0$

Either $(2x + 3) = 0$, giving $x = -\dfrac{3}{2}$ or $(\dots\dots\dots) = 0$, giving $x = \dots\dots$

Substitute each value of x into the linear equation to find the corresponding value of y.

When $x = -\dfrac{3}{2}$, $y = 2(-\dfrac{3}{2}) + 3 = 0$

When $x = \dots\dots$, $y = 2(\dots\dots) + 3 = \dots\dots$

Write down each solution pair.

So the solutions are $x = -\dfrac{3}{2}$, $y = 0$ and $x = \dots\dots$, $y = \dots\dots$

b Rearrange the linear equation to make y the subject.

$$5x - y + 13 = 0$$
$$y = 5x + 13$$

Substitute the linear expression for y into the circle equation.

$$x^2 + (5x + 13)^2 = 13$$

Simplify and rearrange the equation so the right-hand side is 0.

$$x^2 + 25x^2 + 130x + 169 = 13$$
$$26x^2 + 130x + 156 = 0$$
$$x^2 + 5x + 6 = 0$$

Solve the quadratic equation.

$(x + 2)(\dots\dots\dots) = 0$

Either $(x + 2) = 0$, giving $x = -2$ or $(\dots\dots\dots) = 0$, giving $x = \dots\dots$

Substitute each value of x into the linear equation to find the corresponding value of y.

When $x = -2$, $y = 5(-2) + 13 = 3$

When $x = \dots\dots$, $y = 5(\dots\dots) + 13 = \dots\dots$

So the solutions are $x = -2$, $y = 3$ and $x = \dots\dots$, $y = \dots\dots$

① Solve these simultaneous equations.

a $y = x + 3$
$y = x^2 - 4x - 11$

b $y = 2x$
$y = x^2 - 5$

c $y = x - 2$
$y = x^2 + 4x - 6$

....

....

d $y = 3 - x$
$y = 2x^2 - 4x$

e $y = 2 - x$
$y = 3x^2$

f $y = 4x + 3$
$y = 4x^2 - x - 3$

....

....

② Solve these simultaneous equations.   **Hint** Rearrange $x + y = 2$ to make y the subject.

a $x + y = 2$
$y = x^2 - 28$

b $3x + y - 2 = 0$
$y = 3x^2 + 13x - 10$

c $y + 2x + 2 = 0$
$y - 9x = 5x^2$

....

....

③ Solve these simultaneous equations.

a $4y - x - 2 = 0$
$2y^2 = x - 2 + 5y$

b $6y = 10 - x$
$2y^2 = x - 5y + 30$

c $5y = x - 1$
$6y^2 = 3 + x$

....

....

④ Solve these simultaneous equations.

Give your answers correct to 3 significant figures (s.f.) where appropriate.

Hint
Use the quadratic formula $x = \dfrac{-b \pm \sqrt{b^2 - 4ac}}{2a}$

a $y = 4x^2 + 5x$
$y = 3 - 5x$

b $y - 6x + 1 = 0$
$y - 3x^2 = 0$

c $6y^2 - x = 6y + 1$
$x - 3y = 2$

....

....

⑤ Solve these simultaneous equations.   **Hint** Substitute $x = 7y - 50$ into $x^2 + y^2 = 100$

a $x^2 + y^2 = 25$
$y = x - 1$

b $x^2 + y^2 = 10$
$y = -x - 2$

c $x^2 + y^2 = 100$
$x = 7y - 50$

....

....

d $x^2 + y^2 = 41$
$y = 9 - x$

e $x^2 - y^2 = 45$
$x = y + 9$

f $x^2 + y^2 = 40$
$y = 10 - 2x$

....

....

Unit 4 Simultaneous equations   27

6 Solve these simultaneous equations.

a $x^2 + y^2 = 13$
  $2y = 3x$

b $x^2 + y^2 = 169$
  $5x - 12y = 0$

c $x^2 + y^2 = 65$
  $9x + 7y - 65 = 0$

....

....

d $x^2 + y^2 = 74$
  $6y + x = 37$

e $x^2 + y^2 = 82$
  $x + 9y = 0$

f $x^2 + y^2 = 117$
  $5y + x = 39$

....

....

7 Solve the simultaneous equations.
Give your answers correct to 3 s.f.
where appropriate.

Hint
Use the quadratic formula $x = \dfrac{-b \pm \sqrt{b^2 - 4ac}}{2a}$

a $x^2 + y^2 = 30$
  $y = 2x + 1$

b $x^2 + y^2 = 50$
  $y + 2x = 3$

c $x^2 + y^2 = 9$
  $y = 2 - x$

....

....

Exam-style questions

8 Solve these simultaneous equations.
$y = 3x^2 - 2x - 1$
$y - 2x = 1$

Give your answers correct to 2 d.p.

....

.... (6 marks)

9 Solve these simultaneous equations.
$x^2 + y^2 = 125$
$x + y = 15$

....

.... (6 marks)

10 Solve these simultaneous equations.
$x^2 + y^2 = 50$
$y + 2x = 3$
Give your answers correct to 2 d.p.

....

.... (6 marks)

Reflect Would there ever be only one solution or no solutions to a pair of linear and quadratic
or circle simultaneous equations? Explain your answer.

Practise the methods

Answer this question to check where to start.

Check up

Tick the correct quadratic equation that can be used to solve the simultaneous equations
$y - 2x = 1$ and $x^2 + y^2 = 113$

A ⚪
$5x^2 - 4x - 112 = 0$

B ⚪
$5x^2 + 4x - 112 = 0$

C ⚪
$5x^2 + 2x - 112 = 0$

D ⚪
$5x^2 - 2x - 112 = 0$

If you ticked B finish solving the equations. Then go to Q3.

If you ticked A, C or D go to Q1 for more practice.

---

① For each pair of linear and quadratic simultaneous equations
  i substitute the linear equation into the quadratic equation and simplify
  ii hence solve the simultaneous equations

a $y = 2x$
$y = x^2 + 5x - 4$

b $y = 3x + 2$
$y = x^2 - 5x - 6$

c $x = y - 4$
$y - 3 = 2x^2$

---

② For each pair of linear and circle simultaneous equations
  i substitute the linear equation into the circle equation and simplify
  ii hence solve the simultaneous equations.

a $y = x - 1$
$x^2 + y^2 = 13$

b $y - x = 7$
$x^2 + y^2 = 109$

c $y + 3x + 5 = 0$
$x^2 + y^2 = 5$

Exam-style questions

③ Solve these simultaneous equations.
$y = 2x^2 + 4x - 11$
$y - 3x = 4$

(6 marks)

④ Use a graphical method to estimate solutions to the simultaneous equations

$y + x = 3$ and $x^2 + y^2 = 25$

Problem-solve!

(1) The diagram shows a cabbage patch within an allotment.

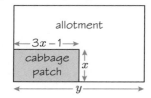

The length of the cabbage patch is half the length of the allotment.

The area of the cabbage patch is 44 m².

Work out the length of the allotment.

....

Exam-style questions

(2) C is the curve with equation $y = x^2 - 5x + 8$
L is the straight line with equation $y = 2x - 4$
L intersects C at two points, A and B.
Calculate the exact length of AB.

.... (6 marks)

(3) A curve with equation $y = x^2 - 2x - 12$ crosses a straight line with equation
$y = 2x + 9$ in two places.
Find the coordinates of the points where they intersect.

.... (6 marks)

(4) The diagram shows a circle of diameter 8 cm with centre at the origin.

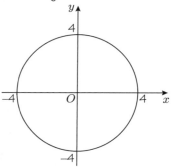

Use an algebraic method to find the points where the line $y = 3x - 1$ crosses the circle.

.... (6 marks)

Now that you have completed this unit, how confident do you feel?

① Solving simultaneous equations
graphically

② Solving linear and non-linear
simultaneous equations algebraically

⑤ Trigonometric graphs

This unit will help you to use and sketch graphs of trigonometric functions.

AO1 Fluency check

① Use your calculator to find these values. Give your answers to 3 significant figures (s.f.)

a cos 20° .... **b** tan 50° .... **c** sin 75° .... **d** tan 80° ....

② **a** Work out the length of AC. Give your answer in surd form.

....

b Complete the table of trigonometric ratios.

sin 30°	sin 60°	cos 30°	cos 60°	tan 30°	tan 60°
....	....	....	....	....	....

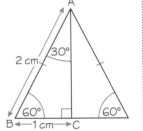

③ **a** Work out the length of PR. Give your answer in surd form.

....

b Complete the table of trigonometric ratios.

sin 45°	cos 45°	tan 45°
....	....	....

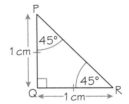

Key points

When using your calculator for trigonometry where angles are measured in degrees, make sure your calculator is set to degrees

The exact values you will be expected to know are sin x and cos x when x = 0°, 30°, 45°, 60° and 90° and tan x when x = 0°, 30°, 45° and 60°.

These **skills boosts** will help you to recognise, interpret and sketch graphs of trigonometric functions.

① Graph of the sine function **②** Graph of the cosine function **③** Graph of the tangent function **④** Solving trigonometric equations

You might have already done some work on trigonometric graphs. Before starting the first skills boost, rate your confidence using each concept.

① Sketch the graph of $y = \sin x$ for $0° \leqslant x \leqslant 360°$. Use your graph to find two values of x when $\sin x = \frac{1}{2}$.

② Sketch the graph of $y = \cos x$ for $0° \leqslant x \leqslant 720°$. Use your graph to find four values of x when $\cos x = \_$.

③ Sketch the graph of $y = \tan x$ for $0° \leqslant x \leqslant 540°$. Use your graph to find three values of x when $\tan x = -1$.

④ Solve $4 \cos x = 0.8$ for $-180° \leqslant x \leqslant 180°$.

How confident are you?

 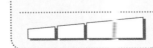 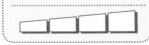

Unit 5 Trigonometric graphs   31

Graph of the sine function

The graph of $y = \sin x$ looks like this.

The graph repeats every 360°.

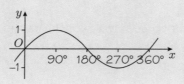

Guided practice

a Sketch the graph of $y = \sin x$ for $0° \leqslant x \leqslant 540°$.

b Use your graph to solve $\sin x = \dfrac{1}{\sqrt{2}}$ for $0° \leqslant x \leqslant 540°$.

---

a Work out how many cycles of the graph you need.

$\dfrac{540}{360} = $ .... cycles

> One cycle is 360°.

Sketch the graph.

Mark the key values on the x- and y-axes.

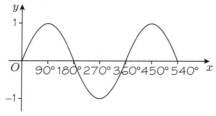

b Work out the value of x between 0° and 90°.

$\sin x = \dfrac{1}{\sqrt{2}}$

$x = 45°$

Draw a line on your sketch at approximately $y = \dfrac{1}{\sqrt{2}}$

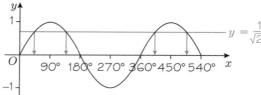

Use the symmetry of the graph to work out the other

values of x in the range that satisfy $\sin x = \dfrac{1}{\sqrt{2}}$

From the graph, there are three other

values for which $\sin x = \dfrac{1}{\sqrt{2}}$

$x = 180° - 45°$

> The graph is symmetrical about $x = 90°$.

$= $ ....

$x = 45° + $ ....

> The graph repeats every 360°.

$= $ ....

$x = 540° - $ ....

> The graph is symmetrical about $x = 450°$.

$= $ ....

List all the values in the range.

When $\sin x = \dfrac{1}{\sqrt{2}}$, $x = 45°, 135°, 405°, 495°$

① Circle the graph of $y = \sin x$ for $-180° \leqslant x \leqslant 180°$.

A

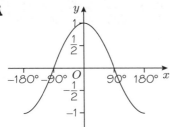

B

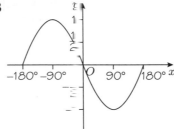

C

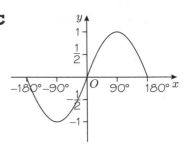

② Use the symmetry properties of the sine graph to complete the following.
Insert numbers between 90 and 360.

a $\sin 0° = \sin$ ....°

b $\sin 240° = \sin$ ....°

c $\sin 330° = \sin$ ....°

d $\sin 75° = \sin$ ....°

③ **a** Sketch the graph of $y = \sin x$ for $0° \leqslant x \leqslant 720°$.

b Use your sketch to work out

i $\sin 630°$

ii $\sin 510°$

iii $\sin 690°$

....

....

....

c Use your graph to solve $\sin x = \dfrac{\sqrt{3}}{2}$ for $0° \leqslant x \leqslant 720°$.

....

④ **a** Draw the graph of $y = \sin x$ for $-360° \leqslant x \leqslant 360°$.

b Use your graph to solve $\sin x = \dfrac{1}{2}$ for $-360° \leqslant x \leqslant 360°$.

....

Exam-style question

⑤ Here is a sketch of $y = \sin x$

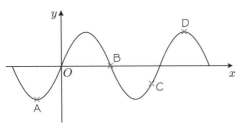

Write down the coordinates of each of the labelled points.

....

(4 marks)

Reflect   Describe the symmetry of the curve $y = \sin x$

2 Graph of the cosine function

The graph of $y = \cos x$ looks like this.

The graph repeats every 360°.

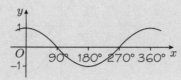

Guided practice

a Sketch the graph of $y = \cos x$ for $0° \leqslant x \leqslant 720°$.

b Use your graph to solve $\cos x = \dfrac{\sqrt{3}}{2}$ for $0° \leqslant x \leqslant 720°$.

---

a Work out how many cycles of the graph you need.

$\dfrac{720}{360} = $ .... cycles

> One cycle is 360°.

Sketch the graph.
Mark the key values on the x- and y-axes.

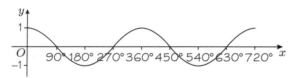

b Work out the value of x between 0° and 90°.

$\cos x = \dfrac{\sqrt{3}}{2}$

$x = 30°$

Draw a line on your sketch at approximately $y = \dfrac{\sqrt{3}}{2}$

Use the symmetry of the graph to work out the other

values of x in the range that satisfy $\cos x = \dfrac{\sqrt{3}}{2}$

From the graph, there are three other values

for which $\cos x = \dfrac{\sqrt{3}}{2}$

$x = 360° - 30°$

> The graph is symmetrical about $x = 180°$.

$\quad = $ ....

$x = 30° + $ ....

> The graph repeats every 360°.

$\quad = $ ....

$x = 720° - $ ....

> The graph is symmetrical about $x = 540°$.

$\quad = $ ....

List all the values in the range.

When $\cos x = \dfrac{\sqrt{3}}{2}$, $x = 30°, 330°, 390°, 690°$

① Circle the graph of $y = \cos x°$ for $-180° \leqslant x \leqslant 180°$.

A
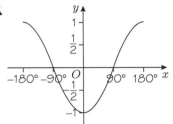

B
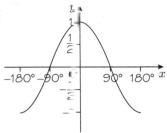

C
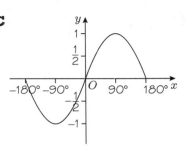

② Use the symmetry properties of the cosine graph to complete the following.

Insert numbers between 90 and 360.

a $\cos 90° = \cos$ ....°   **b** $\cos 120° = \cos$ ....°

c $\cos 30° = \cos$ ....°   **d** $\cos 75° = \cos$ ....°

③ **a** Sketch the graph of $y = \cos x$ for $0° \leqslant x \leqslant 1080°$.

b Use your sketch to work out
   i $\cos 540°$ **ii** $\cos 600°$ **iii** $\cos 1020°$

   ....

c Use your graph to solve $\cos x = \dfrac{1}{2}$ for $0° \leqslant x \leqslant 1080°$.

....

④ **a** Draw the graph of $y = \cos x$ for $-360° \leqslant x \leqslant 360°$.

b Use your graph to solve $\cos x = \dfrac{1}{\sqrt{2}}$ for $-360° \leqslant x \leqslant 360°$.

....

Exam-style question

⑤ Here is a sketch of $y = \cos x$

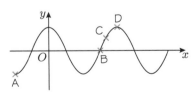

Write down the coordinates of each of the labelled points.

.... (4 marks)

Reflect   Describe the similarities and differences between the curves of $y = \sin x$ and $y = \cos x$

3 Graph of the tangent function

The graph of $y = \tan x$ looks like this.

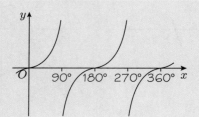

The graph repeats every 180°.

$y = \tan x$ is not defined for 90°, 270°, 450° ...

Guided practice

a Sketch the graph of $y = \tan x$ for $0° \leqslant x \leqslant 540°$.

b Use your graph to solve $\tan x = \dfrac{1}{\sqrt{3}}$ for $0° \leqslant x \leqslant 540°$.

---

a Work out how many cycles of the graph you need.

$\dfrac{540}{180} = $ .... cycles

One cycle is 180°.

Sketch the graph.

Mark the key values on the x-axis.

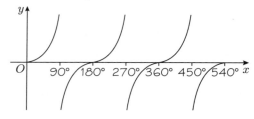

b Work out the value of x between 0° and 90°.

$\tan x = \dfrac{1}{\sqrt{3}}$

$x = 30°$

Draw a line on your sketch to represent $y = \dfrac{1}{\sqrt{3}}$

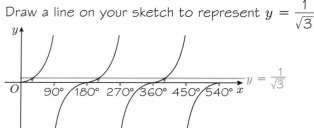

Use the symmetry of the graph to work out the other values of x in the range that satisfy $\tan x = \dfrac{1}{\sqrt{3}}$

From the graph, there are two other values for which $\tan x = \dfrac{1}{\sqrt{3}}$

$x = 30° + 180°$

The graph repeats every 180°.

$\quad = $ ....

$x = 30° + $ ....

$\quad = $ ....

List all the values in the range.

When $\tan x = \dfrac{1}{\sqrt{3}}$, $x = 30°, 210°, 390°$

① Circle the graph of $y = \tan x$ for $-180° \leqslant x \leqslant 180°$.

A 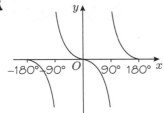 **B** 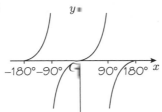 **C**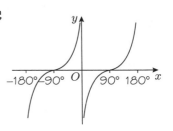

② Use your graph from the guided practice question to complete the following.

Insert numbers between 90 and 360.

a $\tan 0° = \tan$ ....°   **b** $\tan 45° = \tan$ ....°

c $\tan 120° = \tan$ ....°   **d** $\tan 105° = \tan$ ....°

③ **a** Sketch the graph of $y = \tan x$ for $0° \leqslant x \leqslant 720°$.

b Use your sketch to work out
     i $\tan 720°$ **ii** $\tan 495°$ **iii** $\tan 585°$

....

c Use your graph to solve $\tan x = 1$ for $0° \leqslant x \leqslant 720°$.

....

④ **a** Draw the graph of $y = \tan x$ for $0° \leqslant x \leqslant 540°$.

b Use your graph to solve $\tan x = 1$ for $0° \leqslant x \leqslant 540°$.

....

Exam-style question

⑤ **a** Sketch the graph of $y = \tan x$
     in the interval $-360° \leqslant x \leqslant 360°$.

(2 marks)

b $\tan 60° = \sqrt{3}$
     Find the other values in the interval $-360° \leqslant x \leqslant 360°$ for which $\tan x = \sqrt{3}$

....

(2 marks)

Reflect   Explain why it is impossible to work out $\tan 90°$.

4 Solving trigonometric equations

Guided practice

Solve the equation $4 \sin x = 3$ for all values of x in the interval $0° \leqslant x \leqslant 540°$.

$4 \sin x = 3$

Divide both sides of the equation by 4.

$\sin x = \dfrac{3}{4}$

Use $\sin^{-1}$ on your calculator to find one value of x.

$x = \sin^{-1}\left(\dfrac{3}{4}\right) = 48.6°$ (to 3 s.f.)

Sketch the graph of $y = \sin x$ for the interval $0° \leqslant x \leqslant 540°$.

Use the graph to find the other values of x.

From the graph, the other values of x are:

$180° - 48.6° = \dots\dots\dots\dots\dots$

$360° + 48.6° = \dots\dots\dots\dots\dots$

$\dots\dots\dots\dots\dots - 48.6° = \dots\dots\dots\dots\dots$

$x = 48.6°, 131.4°, 408.6°, 491.4°$

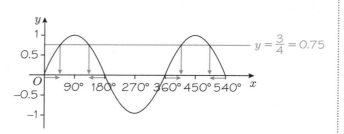

① Solve the equation $5 \sin x = 2$ for all values of x in the interval $0° \leqslant x \leqslant 720°$.

....

② Solve the equation $8 \cos x = 5$ for all values of x in the interval $-360° \leqslant x \leqslant 360°$.

....

③ Solve the equation $7 \tan x = 10$ for all values of x in the interval $0° \leqslant x \leqslant 540°$.

....

④ Solve the equation $5 \tan x = 12$ for all values of x in the interval $0° \leqslant x \leqslant 720°$.

....

Exam-style question

⑤ **a** Sketch the graph of $y = \cos x$ in the interval $-360° \leqslant x \leqslant 360°$.

(2 marks)

b Solve the equation $6 \cos x = 5$ in the interval $-360° \leqslant x \leqslant 360°$.

.... (2 marks)

Reflect

$2 \cos^2 x + 3 \cos x - 2 = (2 \cos x - 1)(\cos x + 2)$

How many solutions are there to $2 \cos^2 x + 3 \cos x - 2 = 0$ in the interval $0° \leqslant x \leqslant 360°$?

Practise the methods

Answer this question to check where to start.

Check up

Here are three equations of graphs.

1 $y = \sin x$ **2** $y = \cos x$ **3** $y = \tan x$

Here are six graphs.

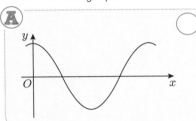
A

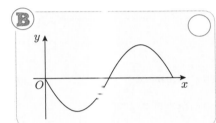
B

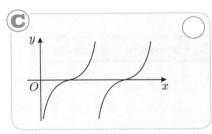
C

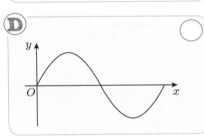
D

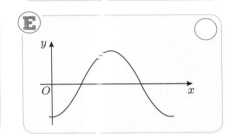
E

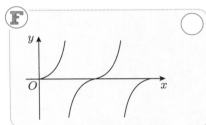
F

Match each equation to the correct graph.

If you matched 1 with D, 2 with A and 3 with F, go to Q2.

If you matched them differently go to Q1 for more practice.

---

(1) This is the graph of $y = \cos x$

Complete the labels on the axes.

180°

Exam-style question

(2) Here is a sketch of $y = \sin x$.

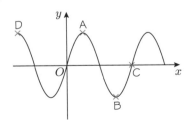

Write down the coordinates of each of the labelled points.

                                                           (4 marks)

(3) Solve the equation $4 \tan x = 15$ for all values of x in the interval $0° \leqslant x \leqslant 720°$.

Problem-solve!

(1) Write down four values of x such that

a $\sin x = 1$

b $\sin x = -\dfrac{1}{2}$

c $\cos x = -1$

....

....

....

d $\cos x = -\dfrac{1}{2}$

e $\tan x = -1$

f $\tan x = \sqrt{3}$

....

....

....

Exam-style question

(2) The diagram shows a sketch of the graph of $y = \sin x$

On the same diagram, draw a sketch of the graph of $y = 2 \sin x$

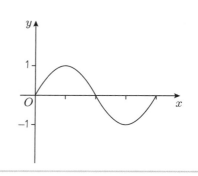

(1 mark)

---

(3) Solve the equation $6 \cos x + 1 = 0$ for all values of x in the interval $0° \leqslant x \leqslant 360°$.

....

(4) Solve the equation $12 \sin x + 5 = 0$ for all values of x in the interval $-360° \leqslant x \leqslant 360°$.

....

(5) Solve the equation $\tan x = -\dfrac{\sqrt{3}}{3}$ for all values of x in the interval $-180° \leqslant x \leqslant 180°$.

....

(6) Solve the equation $\sin (x + 10) = -\dfrac{\sqrt{3}}{2}$ for all values of x in the interval $0° \leqslant x \leqslant 360°$.

....

Now that you have completed this unit, how confident do you feel?

 1 Graph of the sine function

 2 Graph of the cosine function

 3 Graph of the tangent function

 4 Solving trigonometric equations

⑥ Functions

This unit will help you to find inverse and composite functions.

AO1 Fluency check

① Make x the subject of

a $y = 3x + 7$

b $y = \dfrac{2x - 1}{3}$

c $y = 4(x + 3) - 5$

....

....

....

② Write y in terms of t.

a $y = 3x$ and $x = 5t$ ....

b $y = x^2$ and $x = 3t$ ....

c $y = x^2 + x$ and $x = t$ ....

Key points

A function is a rule using values of x to work out values of y.

The function of x is written as f(x) or $x \to \square$. The function f(x) = $2x + 1$ can also be written as $x \to 2x + 1$

These **skills boosts** will help you to use function notation and find inverse and composite functions.

| 1 Using function notation | 2 Inverse functions | 3 Composite functions |

You might have already done some work on functions. Before starting the first skills boost, rate your confidence using each concept.

①
f(x) = $3x + 1$ and
g(x) = $2x^2 - 3$. Write

a 2f(x)

....

b g($2x$)

....

c f(x) + g(x)

....

②
f(x) = $3(x + 2)$

Find f^{-1}(x).

③
f(x) = $2x - 5$ and
g(x) = $x^2 + 3$

Work out gf(x).

How confident are you?

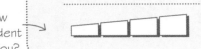

1 Using function notation

2f(x) means double the whole function. For example, for f(x) = 2x + 1, 2f(x) = 2(2x + 1) = 4x + 2
f(2x) means substitute 2x for x in the function.
For example, for f(x) = 2x + 1, f(2x) = 2(2x) + 1 = 4x + 1

Guided practice

f(x) = 4x − 3 and g(x) = 3x² + 5. Write out in full

a 3f(x)

b g(−2x)

c f(x) + g(x)

_____

a Multiply f(x) by 3.

3f(x) = (.....................) ....

| 3f(x) = 3 × f(x) |

= 12x − 9

b Substitute −2x for x in g(x).

g(−2x) = 3(.............)········ + 5

= 12x² + 5

c Add the functions f(x) and g(x) together and then simplify where possible.

f(x) + g(x) = 4x − 3 + ...

= 3x² + 4x + 2

① f(x) = 3x + 4, g(x) = x² + 2x and h(x) = 2x³. Work out   **Hint**   In **a**, substitute x = 4 in f(x).

a f(4)   **b** f(−2)   **c** g(1)   **d** g(−3)

....

e h(3)   **f** h(−1)   **g** f(10) + g(−2)   **h** g(5) − h(2)

....

② f(x) = 4x + 7. Write out in full

a f(x) + 3   **b** f(x) − 5   **c** 2f(x)

....

d 4f(x)   **e** f(3x)   **f** f(5x)

....

③ g(x) = 5x − 1. Write out in full   **Hint**   In **e**, substitute x + 2 for x in g(x).

a g(x) + 4   **b** g(x) − 3   **c** 3g(x)

....

d g(2x)   **e** g(x + 2)   **f** g(−x)

....

(4) $f(x) = x^2 + 3$. Write out in full

a $f(x) + 1$

b $f(x) - 2$

c $2f(x)$

....

....

....

d $f(2x)$

e $f(-x)$

f $f(x + 1)$

....

....

....

(5) $f(x) = 2x^2 - 5$. Write out in full

a $f(x) + 3$

b $f(x) - 1$

c $3f(x)$

....

....

....

d $f(3x)$

e $f(x + 2)$

f $f(x - 1)$

....

....

....

Exam-style question

(6) $f(x) = 2x^2 - 3x - 20$

Express $f(x + 4)$ in the form $ax^2 + bx$

.... **(3 marks)**

(7) Work out $f(x) + g(x)$ when

a $f(x) = 4x + 7$ and $g(x) = 3x - 2$

b $f(x) = 3x + 2$ and $g(x) = x - 5$

....

....

c $f(x) = x + 10$ and $g(x) = 2x - 4$

d $f(x) = x - 5$ and $g(x) = x^2 + 1$

....

....

e $f(x) = 2x + 5$ and $g(x) = 2x^2 + x - 4$

....

(8) $f(x) = 3x + 2$ and $g(x) = x + 4$

a Work out $f(x) - g(x)$.

b Work out $g(x) - f(x)$.

....

....

(9) $f(x) = x + 1$ and $g(x) = 4x - 3$

a Work out $f(x) - g(x)$.

b Work out $g(x) - f(x)$.

....

....

(10) Work out $g(x) - f(x)$ when

a $f(x) = x - 2$ and $g(x) = x^2 + 5$

b $f(x) = 4x^2 + 3x - 1$ and $g(x) = 2x + 3$

....

....

Reflect In Q4, $f(x) = x^2 + 3$ and $f(-x) = x^2 + 3$. Explain why $f(x) = f(-x)$ for this function.
Is $f(x) = f(-x)$ always true for any function? Explain your answer.

2 Inverse functions

The inverse function, written as $f^{-1}(x)$, reverses the function $f(x)$.

Guided practice

Find the inverse of each function.

a $x \to 2x + 5$ **b** $f(x) = \dfrac{2x}{3} - 1$

---

a Put y equal to the function. $y = 2x + 5$

Rearrange to make x the subject.

$2x = y -$ ....

$x = \dfrac{y - ....}{........................}$

Replace y with x.

The inverse function of $x \to 2x + 5$ is $x \to \dfrac{x - 5}{2}$

b Put y equal to $f(x)$.

$y = \dfrac{2x}{3} - 1$

Rearrange to make x the subject.

$y + 1 = \dfrac{2x}{3}$

........($y + 1$) $= 2x$

$x = \dfrac{........................}{........................}$

Replace x with $f^{-1}(x)$ and replace y with x.

$f^{-1}(x) = \dfrac{3(y + 1)}{2}$

Replace y with x.

$f^{-1}(x) = \dfrac{3(x + 1)}{2}$

Rearranging the equation to make x the subject is an alternative method to drawing the inverse function machine.

$x \to \boxed{\times 2} \to \boxed{+5} \to 2x +$

$\boxed{} \leftarrow \boxed{\div 2} \leftarrow \boxed{-5} \leftarrow x$

(1) Find the inverse of each function.

a $x \to 4x + 3$ **b** $x \to 3(x - 5)$ **c** $x \to \dfrac{x}{4} + 1$ **d** $x \to \dfrac{x - 2}{5}$

....

(2) Find the inverse of each function.   **Hint** In **b** simplify the function first.

a $f(x) = 2(x + 3)$ **b** $g(x) = 3(x - 1) + 6$ **c** $h(x) = \dfrac{4x}{3} - 5$

....

Exam-style question

(3) The function f is such that $f(x) = 3x - 1$
Find $f^{-1}(x)$.

.... **(2 marks)**

Reflect Explain why rearranging the equation and finding the inverse function machine give the same answer.

3 Composite functions

To work out a composite function, fg(x), substitute g(x) in f(x).

Guided practice

f(x) = $3 - 2x$ and g(x) = $x^2 + 5$. Work out

a fg(6) **b** fg(x) **c** gf(x)

---

a Work out g(6).

g(6) = ...$^2 + 5$

= ....

Substitute your answer for g(6) into f(x).

fg(6) = $3 - 2($...................$)$

= -79

c Substitute f(x) into g(x).

gf(x) = $($...................$)^2 + 5$

Expand the brackets and simplify.

= .... $+ 5$

= $4x^2 - 12x + 14$

b Substitute g(x) into f(x).

fg(x) = f(g(x))

= $3 - 2($...................$)$

Expand the brackets and simplify.

= 3 ....

= $-2x^2 - 7$

---

(1) f(x) = $x + 2$ and g(x) = $7 - 2x$. Work out

a fg(2) .... **b** gf(5) .... **c** fg(3) ....

(2) f(x) = $2x + 5$ and g(x) = $8 - 5x$. Work out

a fg(x) .... **b** gf(x) .... **c** ff(x) ....

(3) f(x) = $3x + 1$ and g(x) = $x^2 + 2$. Work out

a fg(x) .... **b** gf(x) .... **c** ff(x) ....

(4) f(x) = $5 - x$, g(x) = $4x + 3$ and h(x) = $2x^2 - 9$. Work out

a fg(x) .... **b** gf(x) .... **c** fh(x) ....

d hg(x) .... **e** hf(x) .... **f** gh(x) ....

(5) f(x) = $4x - 1$, g(x) = $3 - x$ and h(x) = $2 - x^2$. Work out

a fg(x) .... **b** gf(x) .... **c** fh(x) ....

d hg(x) .... **e** hf(x) .... **f** gh(x) ....

Exam-style question

(6) f(x) = $2(x - 3)$

Show that ff(x) = $4x - 18$

.... **(2 marks)**

Reflect Can you write any functions f(x) and g(x), where fg(x) = gf(x)?

Practise the methods

Answer this question to check where to start.

Check up

$f(x) = \dfrac{5x}{7} - 4$. Tick the correct function for $f^{-1}(x)$.

A $\quad f^{-1}(x) = \dfrac{7x}{5} + 4$ ◯

B $\quad f^{-1}(x) = \dfrac{7(x + 4)}{5}$ ◯

C $\quad f^{-1}(x) = \dfrac{5x}{7} + 4$ ◯

D $\quad f^{-1}(x) = \dfrac{5(x + 4)}{7}$ ◯

E $\quad f^{-1}(x) = 7\left(\dfrac{x}{5} + 4\right)$ ◯

F $\quad f^{-1}(x) = 7\left(\dfrac{x + 4}{5}\right)$ ◯

If you ticked B or F go to Q3.　　　　　If you ticked A, C, D or E go to Q1 for more practice.

1 Find $f^{-1}(x)$ when $f(x) = \dfrac{5x}{7} - 4$

$$y = \dfrac{5x}{7} - 4$$

$+4 \left(\qquad\qquad \right) +4$
$= $
$\times 7 \left(\qquad\qquad \right) \times 7$
$= $
$\div 5 \left(\qquad\qquad \right) \div 5$
$= $

$f^{-1}(x) = $ ....

2 Find $f^{-1}(x)$ for each function.

a $f(x) = 3x - 7$　　　**b** $f(x) = 4(x + 9)$　　　**c** $f(x) = \dfrac{x - 8}{3}$　　　**d** $f(x) = 2 + \dfrac{x}{5}$

Exam-style question

3 The function f is such that $f(x) = 5x + 9$
Find $f^{-1}(x)$.

.... (2 marks)

4 $f(x) = 6 - 5x$ and $g(x) = 2x^2 - 5$. Write out in full

a $f(x) + 2$　　　　**b** $2f(x)$　　　　**c** $f(3x)$

d $f(x + 2)$　　　　**e** $g(x - 1)$　　　　**f** $f(x) + g(x)$

Exam-style questions

5 $f(x) = 3x^2 - x + 4$
Express $f(x + 2)$ in the form $ax^2 + bx + c$

.... (3 marks)

6 $f(x) = 3(x + 1)$ and $g(x) = 3(x - 2)$
Show that $fg(x) = 9x - 15$

....

.... (2 marks)

Problem-solve!

① $f(x) = 5(x - 1)$ and $g(x) = 5(x + 1)$
Work out the value of a when $f^{-1}(a) + g^{-1}(a) = 1$

....

Exam-style questions

② The functions f and g are such that $f(x) = \; \cdot \; - 3x$ and $g(x) = 3x - 1$
   a Show that $gf(1) = -7$

   .... **(2 marks)**

   b Prove that $f^{-1}(x) + g^{-1}(x) = \dfrac{2}{3}$ for all values of x.

   .... **(3 marks)**

③ The function f is such that $f(x) = 2x - 1$
   a Find $f^{-1}(x)$.

   .... **(2 marks)**

   The function g is such that $g(x) = kx^2$ where k is a constant.
   b Given that $fg(2) = 39$ work out the value of k.

   .... **(2 marks)**

④ $f(x) = x - 5$ and $g(x) = x^2 + 8$
   a Work out
      i $fg(x)$ .... **(2 marks)**

      ii $gf(x)$ .... **(2 marks)**

   b Solve $fg(x) = gf(x)$

   .... **(2 marks)**

⑤ $f(x) = \dfrac{x}{3} - 4$ and $g(x) = 3(x + 4)$

   a Work out
      i $fg(x)$ **ii** $gf(x)$

   ....

   b Are $f(x)$ and $g(x)$ inverse functions?
   Explain your answer.

   Hint $f(x)$ and $g(x)$ are inverse functions if $fg(x) = gf(x) = x$

   ....

⑥ Are $f(x) = \dfrac{x + 5}{6}$ and $g(x) = 6(x - 5)$ inverse functions? Explain your answer.

....

Now that you have completed this unit, how confident do you feel?

Transformations of graphs

This unit will help you to transform graphs and interpret the transformations of graphs.

① Sketch each graph, labelling the points where the graph crosses the axes.

   a $y = (x + 2)(x - 3)$ **b** $y = (x + 1)(x - 2)^2$ **c** $y = \sin x$ **d** $y = \cos x$

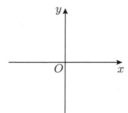 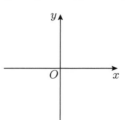  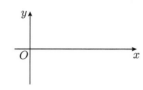

② **a** The point A(3, −4) is reflected in the x-axis to point B.

     Write down the coordinates of B.   ....

   b The point P(−1, 2) is translated by $\begin{pmatrix} 5 \\ -4 \end{pmatrix}$ to point Q.

     Write down the coordinates of Q.   ....

Key points
↓

Graphs can be translated, reflected or stretched.

These **skills boosts** will help you to translate, reflect and stretch graphs, interpret the transformations of graphs and write functions algebraically.

1 Transforming graphs ⟩ **2 Interpreting transformations of graphs**

You might have already done some work on transforming graphs. Before starting the first skills boost, rate your confidence using each concept.

① This is the graph of $y = f(x)$

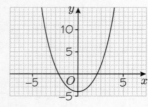

On the same grid
a sketch and label the graph of
   $y = f(x + 3)$
b sketch and label the graph of
   $y = -f(x)$

② $y = f(x)$ where $f(x) = x^2$
Graph A is a translation of $f(x)$

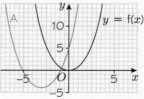

Work out the equation of graph A
in terms of x.

....

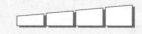

1 Transforming graphs

$y = f(x) + a$ is a vertical translation of $y = f(x)$ by a.
$y = f(x + a)$ is a horizontal translation of $y = f(x)$ by $-a$.
$y = f(-x)$ is a reflection of $y = f(x)$ in the y-axis.
$y = -f(x)$ is a reflection of $y = f(x)$ in the x-axis.
$y = af(x)$ is a vertical stretch of $y = f(x)$ of scale factor a, so multiply the y-coordinates by a.
$y = f(ax)$ is a horizontal stretch of $y = f(x)$ of scale factor $\frac{1}{a}$, so multiply the x-coordinates by $\frac{1}{a}$.

Guided practice

This is the graph of $y = f(x)$
Sketch the graphs of

a i $y = f(x + 3)$ **ii** $y = f(x) + 2$
b i $y = -f(x)$ **ii** $y = f(-x)$
c i $y = 2f(x)$ **ii** $y = f(2x)$

a i To draw the graph of
$y = f(x + 3)$, translate
$y = f(x)$ horizontally by -3.
Label the curve $y = f(x + 3)$

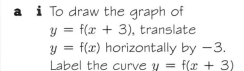

Translating horizontally
by -3 means translating
3 units to the left.

ii To draw the graph of $y = f(x) + 2$,
translate $y = f(x)$ vertically by 2.
Label the curve $y = f(x) + 2$

Translating vertically
by 2 means translating
2 units up.

b i To draw the graph of
$y = -f(x)$, reflect
$y = f(x)$ in the x-axis.
Label the curve
$y = -f(x)$

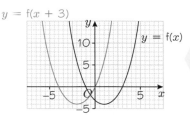

ii To draw the graph of
$y = f(-x)$ reflect
$y = f(x)$ in the y-axis.
Label the curve $y = f(-x)$

c i To draw the graph of
$y = 2f(x)$, vertically
stretch $y = f(x)$ by
scale factor 2.
Label the curve
$y = 2f(x)$

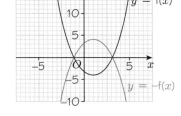

Multiply all the
y-coordinates by 2.

ii To draw the graph of $y = f(2x)$,
horizontally stretch $y = f(x)$ by
scale factor $\frac{1}{2}$.
Label your curve $y = f(2x)$

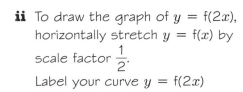

Multiply all the
x-coordinates by $\frac{1}{2}$.

① The graph of $y = f(x)$ is shown on the grid.

On the same grid, sketch and label the graphs of

a $y = f(x + 5)$

b $y = f(x) - 4$

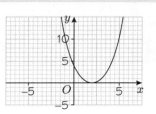

② The graph of $y = f(x)$ is shown on the grid.

On the same grid, sketch and label the graphs of

a $y = f(-x)$

b $y = -f(x)$

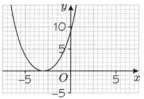

③ The graph of $y = f(x)$ is shown on the grid.

On the same grid, sketch and label the graphs of

a $y = 2f(x)$

b $y = f(2x)$

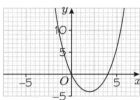

Exam-style questions

④ The graph of $y = f(x)$ is shown on both grids.
a On this grid, sketch the graph of $y = f(x - 3)$ **b** On this grid, sketch the graph of $y = 2f(x)$

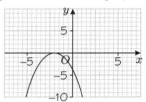

(2 marks)

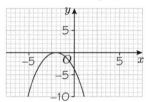

(2 marks)

⑤ The graph of $y = f(x)$ is shown on both grids.
a On this grid, draw the graph of $y = -f(x)$ **b** On this grid, draw the graph of $y = f(x + 2)$

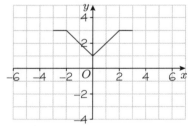

(1 mark)

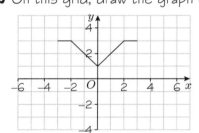

(1 mark)

⑥ **a** Here is the graph of $y = \cos x$
for $-180° \leqslant x \leqslant 180°$.

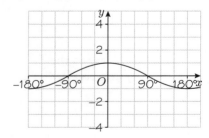

On this grid, sketch the graph of
$y = \cos x - 2$ for $-180° \leqslant x \leqslant 180°$.

(2 marks)

b Here is the graph of $y = \sin x$
for $-180° \leqslant x \leqslant 180°$.

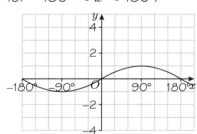

On this grid, sketch the graph of
$y = 3 \sin x$ for $-180° \leqslant x \leqslant 180°$.

(2 marks)

Reflect   Use the graphs in Q6 to describe the graph of $y = \sin x$ as a transformation
of $y = f(x) = \cos x$.

2 Interpreting transformations of graphs

$y = f(x)$ where $f(x) = x^2$

Graph A is a translation of $f(x)$.

Find the equation of graph A in terms of x.

_____

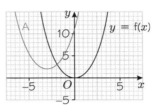

$f(x) = x^2$

Decide whether the transformation is a translation, a reflection or a stretch.

The graph is the same but moved to the left and up, so the transformation is a ....

Work out what $f(x)$ is translated by.

Graph A is a translation of $f(x)$ by .... units left

and .... units up.

Apply this transformation to $f(x)$.

Graph A $= f(x + 3) + 2$

Substitute $(x + 3)$ into $f(x) = x^2$ and add 2.

> A translation of $f(x)$ by 3 units left is $f(x + 3)$.
> A translation of 2 units up is $f(x) + 2$.

$= ($................$)^2 +$ ....

Expand and simplify. $= x^2 + 6x + 11$

① $y = f(x)$ where $f(x) = (x - 1)^3$

Graph R is a transformation of $f(x)$.

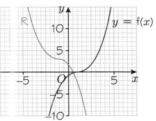

Find the equation of graph R in terms of x.

....

② The graph of $y = f(x)$ is shown on the grid, where $f(x) = (x - 2)^2$

The graph G is a transformation of the graph of $y = f(x)$

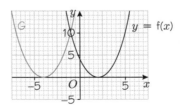

Write down the equation of graph G.

.... (2 marks)

Reflect Looking at Q1 and Q2, which transformation was easier to write the equation for? Why?

Practise the methods

Answer this question to check where to start.

Check up

The diagram on the right shows the graph of $y = f(x) = 2x + 1$
Below, the graph has been transformed in different ways.
Match each graph to the correct function notation.

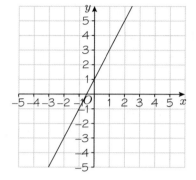

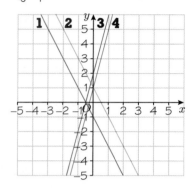

A $y = 2f(x)$ ◯

B $y = f(2x)$ ◯

C $y = -f(x)$ ◯

D $y = f(-x)$ ◯

If you matched 1 with C, 2 with D, 3 with A and 4 with B go to Q3.

If you matched them differently go to Q1 for more practice.

---

① $f(x) = 2x - 3$

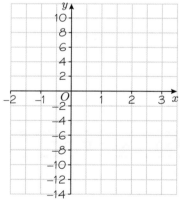

a **i** Complete the table of values for $f(x)$.

x	-2	-1	O	1	2	3
$f(x)$						

ii Sketch and label the graph of $y = f(x)$

b **i** Complete the table of values for $2f(x)$.

x	-2	-1	O	1	2	3
$2f(x)$						

ii Explain the relationship between the coordinates for $f(x)$ and $2f(x)$.

....

iii Sketch and label the graph of $y = 2f(x)$ on the same grid as your sketch of $y = f(x)$

iv Describe how $y = f(x)$ transforms to $y = 2f(x)$

....

c **i** Complete the table of values for f(2x).

x	−2	−1	0	1	2	3
f(2x)						

ii Sketch and label the graph of $y = f(2x)$ on the same grid as your sketch of $y = f(x)$

iii Describe how $y = f(x)$ transforms to $y = f(2x)$

....

(2) Describe the transformation that maps the graph of $y = f(x)$ onto the graph of

a $y = f(x) + 7$

....

b $y = f(x − 3)$

....

c $y = f(−x)$

....

d $y = f(3x)$

....

e $y = 5f(x)$

....

(3) The graph of $y = f(x)$ is shown on the grid.

On the same grid, sketch and label the graphs of

a $y = f(x + 4)$

b $y = f(x) + 2$

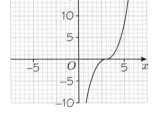

(4) The graph of $y = f(x)$ is shown on the grid.

On the same grid, sketch and label the graphs of

a $y = f(−x)$

b $y = f(2x)$

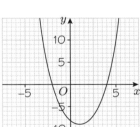

Exam-style question

(5) The graph of $y = f(x)$ is shown on the grid, where $f(x) = x^2 − 3$

The graph P is a transformation of the graph of $y = f(x)$

Write down the equation of graph P.

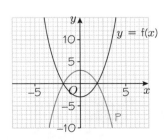

.... **(2 marks)**

Problem-solve!

Exam-style questions

(1) The graph shows $y = g(x)$

The graph G is the reflection of $y = g(x)$ in the x-axis.

Write down the equation of graph G.

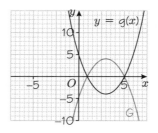

.... **(2 marks)**

(2) The graph of $y = f(x)$ is transformed to give the graph of $y = -f(x - 4)$
The point A on the graph of $y = f(x)$ is mapped to the point P on the graph of $y = -f(x - 4)$
The coordinates of point A are (5, 2).
Find the coordinates of point P.

.... **(2 marks)**

(3) The graph shows $y = f(x)$

On the same grid, sketch the graph of
$y = -f(x) + 3$

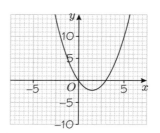

(1 mark)

(4) Here is the graph of $y = \sin x$
for $-180° \leqslant x \leqslant 180°$.

On the same grid, sketch the graph of
$y = -3 \sin x$ for $-180° \leqslant x \leqslant 180°$.

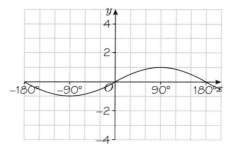

(2 marks)

(5) Here is a sketch of the curve $y = a \cos bx + c$,
$0° \leqslant x \leqslant 360°$.

Find the values of a, b and c.

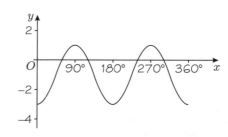

.... **(3 marks)**

Now that you have completed this unit, how confident do you feel?

1 Transforming graphs

2 Interpreting transformations of graphs

8 Pre-calculus

This unit will help you to estimate and interpret the gradient of a non-linear graph and the area under a non-linear graph.

AO1 Fluency check

1 Calculate the area of this trapezium.

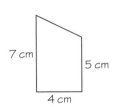

7 cm

5 cm

4 cm

2 Work out the gradient of each line.

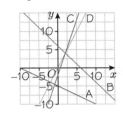

....

Key points

| The gradient of a straight line is $\dfrac{\text{change in } y}{\text{change in } x}$ | The tangent to a curve at a given point is the straight line that touches the curve at that point. | The gradient at a point on a curve is the gradient of the tangent at that point. |

These **skills boosts** will help you to interpret the gradient, estimate the gradient at a point on a curve and estimate the area under a graph.

1 Gradient

2 Area under a curve

You might have already done some work on gradient and area under graphs. Before starting the first skills boost, rate your confidence using each concept.

1 On a distance–time graph, what does the gradient of the tangent to the curve represent?

....

....

....

....

2 Use the graph of $y = -\dfrac{1}{10}x^2 + x + \dfrac{15}{2}$ to estimate the area under the graph of $y = -\dfrac{1}{10}x^2 + x + \dfrac{15}{2}$ from $x = 5$ to $x = 8$.

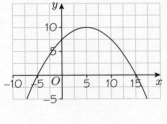

How confident are you?

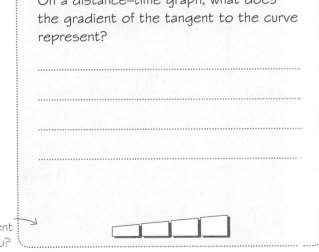

Gradient

The gradient of the tangent at any point on a distance–time graph gives the speed at that point.

The gradient of the tangent at any point on a speed–time graph gives the acceleration at that point.

Guided practice

Lexi runs in a race.

The graph shows her speed, in metres per second, t seconds after the start of the race.

a Calculate an estimate for the gradient of the graph when $t = 4$

b Describe fully what your answer to part **a** represents.

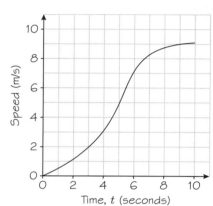

_____

a On the graph, draw a tangent to the curve at the point when $t = 4$

$$\text{Gradient} = \frac{\text{change in speed}}{\text{change in } t}$$

$$= \frac{\text{......} - \text{......}}{\text{......} - \text{......}}$$

$$= 1.4$$

The tangent to a curve at a given point is the straight line that touches the curve at that point. For example,

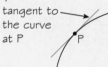

tangent to the curve at P

b 4 seconds after the start of the race,

Lexi was .... at 1.4 m/s².

What does the gradient of the tangent at any point on a speed–time graph represent?

Exam-style question

① The graph shows information about the distance, d metres, of a cyclist t seconds after setting off.

a Work out an estimate for the gradient when $t = 3$

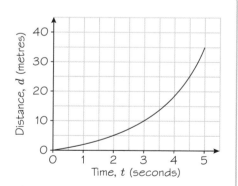

.... (2 marks)

b Write an interpretation of this gradient.

.... (1 mark)

Reflect What would a negative gradient on a speed–time graph represent?

2 Area under a curve

To estimate the area under part of a curve, draw a chord between the two required points and straight lines down to the horizontal axis to form a trapezium. The area of the trapezium is an estimate for the area under this part of the curve.

The area under a speed–time graph is the distance.

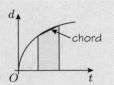

Guided practice

Here is a speed-time graph for a car journey.

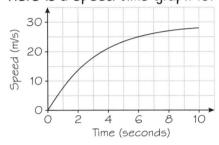

Estimate the distance the car travelled between 4 and 8 seconds.

Draw chords between 4 and 6 seconds and between 6 and 8 seconds.

Draw straight lines down from the ends of the chords to the time axis.

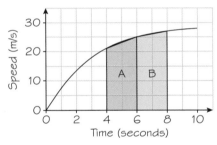

Work out the total area of the trapezia.

Area of trapezium A $= \frac{1}{2} \times 2(21 + \ldots\ldots)$

$\qquad = \ldots\ldots$

Area of trapezium B $= \frac{1}{2} \times \ldots\ldots(\ldots\ldots + \ldots\ldots)$

$\qquad = \ldots\ldots$

Total area $= \ldots\ldots + \ldots\ldots$

$\qquad = 98$

An estimate for the distance the car travelled between 4 and 8 seconds is 98 metres.

The distance the car travelled between 4 and 8 seconds is the area under the curve between 4 and 8 seconds.

The total area of the trapezia gives an estimate for the area under the curve between 4 and 8 seconds.

Area of trapezium $= \frac{1}{2}h(a + b)$

(1) The speed–time graph shows a van driving away from traffic lights.

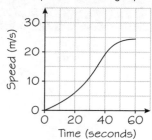

The time after the traffic lights, t, is measured in seconds, s.

The speed, s, is measured in metres per second (m/s).

Estimate the distance travelled between $t = 40$ and $t = 60$.

Use 2 strips of equal width.

....

(2) **a** Draw the graph of $y = x^2 + 4$ for $0 \leqslant x \leqslant 3$.

b Estimate the area under the graph of $y = x^2 + 4$ between $x = 0$ and $x = 3$.

Use 3 strips of equal width.

....

Exam-style question

(3) The graph shows the speed of a plane when preparing to take off.

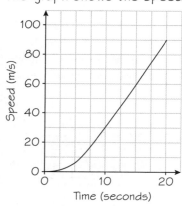

Work out an estimate for the distance travelled by the plane in the first 20 seconds of preparing to take off.

Use 4 strips of equal width.

.... **(3 marks)**

Reflect How could you make your estimates for the area under the curves on this page more accurate?

Practise the methods

Answer this question to check where to start

Check up

Tick the correct calculation for an estimate of the gradient at the point when $x = 2$.

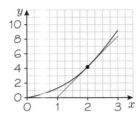

A $\dfrac{3-1}{8-0}$ ○

B $\dfrac{8-0}{3-1}$ ○

C $\dfrac{0-8}{3-1}$ ○

D $\dfrac{8-0}{1-3}$ ○

E $\dfrac{1-3}{8-0}$ ○

F $\dfrac{3-1}{0-8}$ ○

If you ticked B finish calculating the gradient. Then go to Q2.

If you ticked A, C, D, E or F go to Q1 for more practice.

1 Work out the gradient of the tangent to the curve at

a $x = -1$ ....

b $x = 0.5$ ....

c $x = 1.5$ ....

Exam-style question

2 The graph shows the velocity of a ball t seconds after being dropped from a tall building.

a Work out an estimate for the acceleration of the ball at $t = 2$.

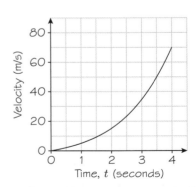

.... (2 marks)

b Work out an estimate for the distance the ball fell in the first 4 seconds after being dropped. Use 4 strips of equal width.

.... (3 marks)

Problem-solve!

(1) Here is a circle, centre O.

Estimate the gradient of the tangent when $x = 3$.

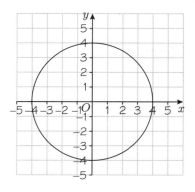

....

(2) The distance–time graph shows information about
a 20 km bike race between Abi and Bella.

a Describe the race between Abi and Bella.

....

....

....

.... **(2 marks)**

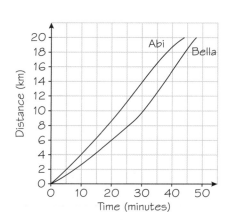

b Estimate the difference between Abi's and Bella's speeds 30 minutes into the race.

.... **(2 marks)**

(3) This is the graph of $y = \frac{1}{2}x\left(6 - \frac{1}{2}x\right)$

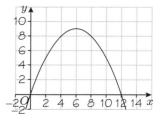

Calculate an estimate for the area bounded
by the curve and the x-axis.

Use 6 strips of equal width.

....

Now that you have completed this unit, how confident do you feel?

1 Gradient

2 Area under a curve

Answers

Unit 1 Graphs

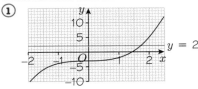

AO1 Fluency check

① **a** v **b** i **c** vi **d** iii **e** ii **f** iv

Confidence check

①

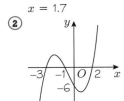

$x = 1.7$

②

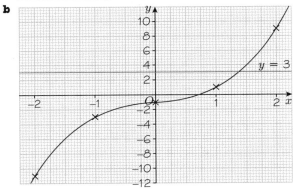

Skills boost 1 Recognising and interpreting graphs of cubic functions

Guided practice

a When $x = -2$, $y = (-2)^3 + (-2) - 1 = \underline{-11}$
When $x = 0$, $y = (0)^3 + (0) - 1 = \underline{-1}$
When $x = 1$, $y = (1)^3 + (1) - 1 = \underline{1}$

x	-2	-1	0	1	2
y	-11	-3	-1	1	9

b

c $x^3 + x - 1 = 3$ when $y = 3$
$x = 1.4$

① C and E.

② **a**

x	-2	-1	0	1	2	3
y	$\underline{5}$	7	$\underline{3}$	$\underline{-1}$	1	$\underline{15}$

b

c $x = -2$, $x = -0.4$ and $x = 2.4$

③ **a**

x	-2	-1	0	1	2
y	-3	$\underline{4}$	$\underline{-1}$	-6	$\underline{1}$

b

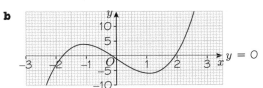

c $x = -1.8$, $x = -0.1$ and $x = 1.9$

④ **a**

x	-2	-1	0	1	2	3
y	$\underline{-3}$	0	$\underline{-3}$	$\underline{-6}$	-3	$\underline{12}$

b

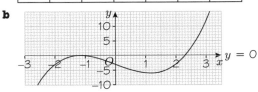

c $x = -1.3$, $x = -1$ and $x = 2.3$

Skills boost 2 Sketching cubic graphs

Guided practice

a $0 = (x - 3)(x - 1)(x + 2)$
$(x - 3) = 0$, so $x = 3$
or $(x - 1) = 0$, so $x = \underline{1}$
or $(x + 2) = 0$, so $x = \underline{-2}$
So the roots of the equation are $x = 3$, $x = \underline{1}$
and $x = \underline{-2}$
The curve crosses the x-axis at the points $(3, 0)$,
$(\underline{1}, \underline{0})$ and $(\underline{-2}, \underline{0})$.
$y = (0 - 3)(0 - 1)(0 + 2) = 6$
The curve crosses the y-axis at $(\underline{0}, \underline{6})$.
$y = (x - 3)(x - 1)(x + 2)$
$\quad = (x - 3)(x^2 + x - 2)$
$\quad = x^3 \underline{- 2x^2 - 5x + 6}$

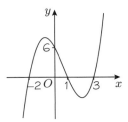

b When $y = 0$, $0 = -(x - 2)^2(x + 1)$
$(x - 2) = 0$, so $x = 2$
or $\underline{(x + 1)} = 0$, so $x = \underline{-1}$
The roots of the equation are $x = 2$ and $x = \underline{-1}$
The curve touches the x-axis at the point $(2, 0)$ and
crosses the x-axis at $(\underline{-1}, \underline{0})$.
$y = -(0 - 2)^2(0 + 1) = \underline{-4}$
So the curve crosses the y-axis at $(\underline{0}, \underline{-4})$.
$y = -(x - 2)^2(x + 1)$
$\quad = -(x + 1)(x^2 - 4x + 4)$
$\quad = -x^3 \underline{+ 3x^2 - 4}$

1 a iv **b** vi **c** ii **d** v **e** i **f** iii

2 a
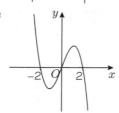

b
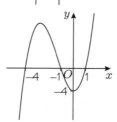

c
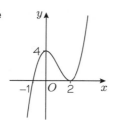

d
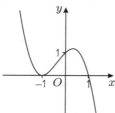

e
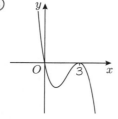

f
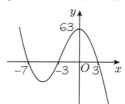

3
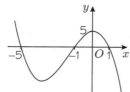

Practise the methods

1 a i $x = -1$, $x = 1$ and $x = -5$
   ii $(-5, 0)$, $(-1, 0)$ and $(1, 0)$
 b $(0, 5)$
 c
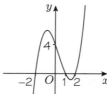

2 a
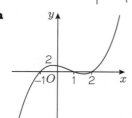

b
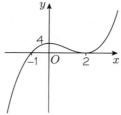

3 a

x	-3	-2	-1	0	1	2	3
y	-4	8	8	2	-4	-4	8

b
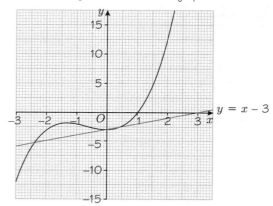

c $x = -2.8$, $x = 0.3$ and $x = 2.5$

Problem-solve!

1 $a = 2$, $b = -11$, $c = -12$
2 $a = 6$, $b = -3$, $c = -10$
3 a **A iv**, **B iii**, **C i** and **D v**
   b $y = -(x + 3)(x - 3)(x + 7)$
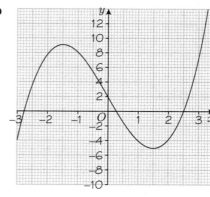

4 Add the line $y = x - 3$ to the graph.
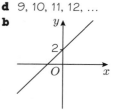

$x = 0$, $x = 0.4$ and $x = -2.4$

Unit 2 Graphical inequalities

1 a 7, 6, 5, 4, …   **b** −3, −2, −1, 0, …
 c −1, 0, 1, 2, …   **d** 9, 10, 11, 12, …
2 a

b
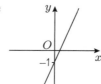

c
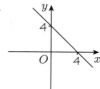
d

Confidence check

(1)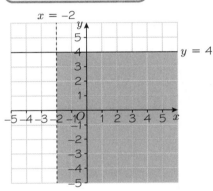

$y = -1$

$x = 3$

(2)

$y = x$

$y = x - 2$

$y = 1$

(3) $x < -3$ or $x > \dfrac{5}{2}$

Skills boost 1 Linear inequalities in one variable

Guided practice

$x = -2$

$y = 4$

(1) a $x \geqslant 0$ and $y > -4$ **b** $x \geqslant -1$ and $y < 2$

(2) $x = -2$

$y = 1$

R

Skills boost 2 Linear inequalities in two variables

Guided practice

At point (1, 1), $x + y = 1 + 1 = \underline{2} < 3$, so the inequality is correct.

At point (1, 0), $y = \underline{0} < x$, so the inequality is incorrect.

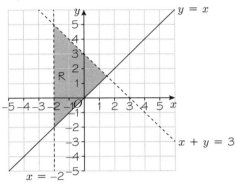

$y = x$

R

$x + y = 3$

$x = -2$

(1) a $y \leqslant x + 2$ **b** $y > 1 - x$

(2) a $x + y \leqslant 3$ and $y < x + 2$

b $x > -4$, $y \geqslant x$ and $x + y < 1$

(3) a

$y = -1$

$x + y = 4$

b

$y = x + 3$

$y = 2 - x$

(4) a

$x = 2$

$y = 2$

$y + x = -2$

b

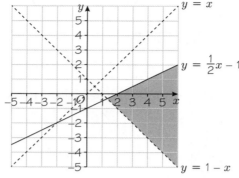

⑤ a

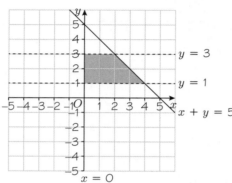

b

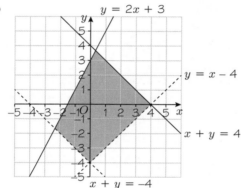

⑥

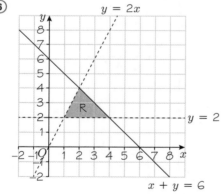

Skills boost 3 Quadratic inequalities

Guided practice

a $y = 2x^2 + 6x - 8$
$\quad = 2(x^2 + 3x - 4)$
$\quad = 2(x + 4)(x - 1)$

When $y = 0$, $0 = 2(x + 4)(x - 1)$
So either $(x + 4) = 0$, giving $x = \underline{-4}$
or $(x - 1) = 0$, giving $x = \underline{1}$
When $x = 0$, $y = 2(0)^2 + 6(0) - 8 = \underline{-8}$

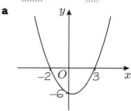

b $x \leqslant \underline{-4}$ or $x \geqslant \underline{1}$

① a

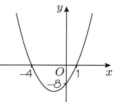

b $-2 < x < 3$

② a

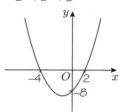

b $x < -4$ or $x > 2$

③ a

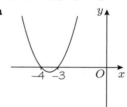

b $x \leqslant -4$ or $x \geqslant -3$

④ $-6 < x < 1$

⑤ a $x < 1$ or $x > 5$ **b** $-4 \leqslant x \leqslant \frac{1}{2}$

   c $x < -5$ or $x > 2$ **d** $x \leqslant -3$ or $x \geqslant 3$

   e $x \leqslant -5$ and $x \geqslant 3$ **f** $3 < x < 4$

⑥ $x < 2$ or $x > 3$

Practise the methods

① a $x < 5$ **b** $y \geqslant -1$ **c** $y > x$

② a

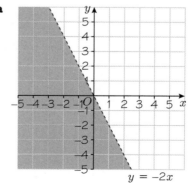

b

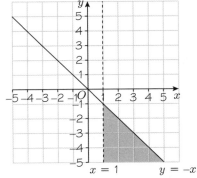

$x = 1$ $y = -x$

c

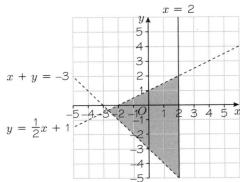

$x = 2$

$x + y = -3$

$y = \frac{1}{2}x + 1$

③ $x \leqslant -2$ or $x \geqslant 6$

Problem-solve!

① 5

② $-4 < x < \frac{2}{3}$

③ $1 \leqslant x \leqslant 4$

④ $x < -1$ or $x > \frac{5}{2}$

⑤ $0 <$ length < 15 cm and $0 <$ width < 7 cm

Unit 3 Exponential graphs

AO1 Fluency check

① **a** v   **b** i   **c** vi   **d** iii   **e** iv   **f** ii

Confidence check

①
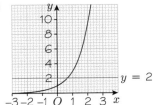

$y = 2$

$x = 0.6$

②
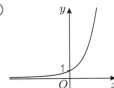

③
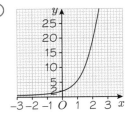

Skills boost 1 Drawing and interpreting exponential graphs of the form $y = a^x$

> **Guided practice**

a When $x = -2$, $y = 2^{-2} = \frac{1}{2^2} = \frac{1}{4}$

When $x = 0$, $y = 2^0 = \underline{1}$

When $x = 1$, $y = 2^1 = \underline{2}$

When $x = 3$, $y = 2^3 = \underline{8}$

x	-3	-2	-1	0	1	2	3
y	$\frac{1}{8}$	$\frac{1}{4}$	$\frac{1}{2}$	1	2	4	8

b
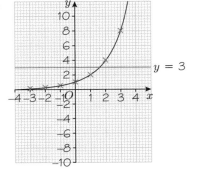

$y = 3$

c $x = 1.6$

① **a**

x	-2	-1	0	1	2	3
y	$\frac{1}{9}$	$\frac{1}{3}$	1	3	9	27

b
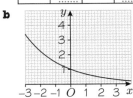

c $x = 1.5$

② **a**

x	-3	-2	-1	0	1	2	3
y	$\frac{27}{8}$	$\frac{9}{4}$	$\frac{3}{2}$	1	$\frac{2}{3}$	$\frac{4}{9}$	$\frac{8}{27}$

b

c $x = -1.7$ **d** $y = 0.5$

③ **a**

x	-2	-1	0	1	2
y	25	5	1	$\frac{1}{5}$	$\frac{1}{25}$

b

c $x = -1.4$

Skills boost 2 Recognising and sketching exponential graphs

Guided practice

When x is large and negative, 2^x is very small.

When x is large and positive, 2^x is very ___large___ .

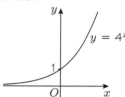

(1) B and F

(2)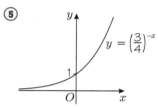
$y = 4^x$

(3)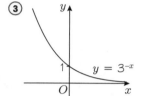
$y = 3^{-x}$

(4)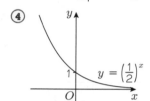
$y = \left(\frac{1}{2}\right)^x$

(5) $y = \left(\frac{3}{4}\right)^{-x}$

(6) A $y = 10^{-x}$

B $y = 10^x$

Skills boost 3 Drawing and interpreting exponential graphs of the form $y = ab^x$

Guided practice

a When $x = -2$, $y = 3(2^2) = $ __12__

When $x = 0$, $y = 3(2^0) = $ __3__

When $x = 1$, $y = 3(2^{-1}) = \dfrac{3}{2}$

When $x = 2$, $y = 3(2^{-2}) = \dfrac{3}{2^2} = \dfrac{3}{4}$

x	-3	-2	-1	0	1	2	3
y	24	12	6	3	$\frac{3}{2}$	$\frac{3}{4}$	$\frac{3}{8}$

b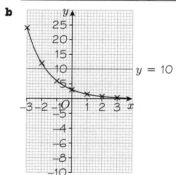
$y = 10$

c $x = -1.7$

(1) a

x	-2	-1	0	1	2
y	36	12	4	$\frac{4}{3}$	$\frac{4}{9}$

b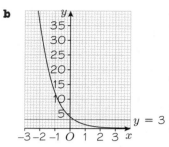
$y = 3$

c $x = 0.3$

Practise the methods

(1) a quadratic

b exponential

c cubic

d linear

(2) a

x	-2	-1	0	1	2	3
y	$\frac{1}{4}$	$\frac{1}{2}$	1	2	4	8

b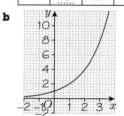

c $x = 2.3$

Problem-solve!

(1) a, b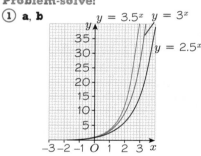
$y = 3.5^x$ $y = 3^x$
$y = 2.5^x$

c $(0, 1)$

(2) a $I = 500 \times 1.02^t$

b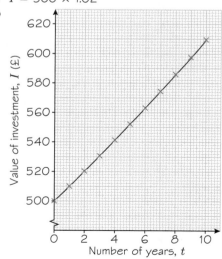

c 9.25 years = 9 years and 3 months

(3) $k = 0.8$ and $a = 5$

Unit 4 Simultaneous equations

AO1 Fluency check

① **a** $y = -2x + 1$

  b $y = 2x + 5$

  c $y = -\frac{1}{2}x + \frac{1}{4}$

② **a** $x^2 + 2x - 15$

  b $2x^2 - 5x - 12$

  c $6x^2 - 13x - 5$

③ **a** $x = -\frac{2}{3}$ or $x = 4$

  b $x = 2$ or $x = 3$

  c $x = -4$ or $x = -1\frac{1}{2}$

Confidence check

① $x = 2.8, y = -0.8$ and $x = -1.8, y = 3.8$

② **a** $x = \frac{2}{5}, y = 3\frac{4}{5}$ and $x = -2, y = 11$

  b $x = 3.11, y = -3.22$ and $x = -0.71, y = 4.42$

Skills boost 1 Solving simultaneous equations graphically

Guided practice

$y + x^2 = 5$

$\quad y = \underline{\mathbf{5 - x^2}}$

$y - 1 = x$

$\quad y = \underline{\mathbf{x + 1}}$

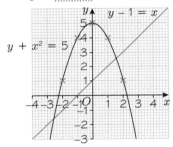

$x = -2.6, y = -1.6$ and $x = \underline{\mathbf{1.6}}, y = \underline{\mathbf{2.6}}$

① $x = -2, y = 0$ and $x = 2.5, y = 2.25$

② **a**

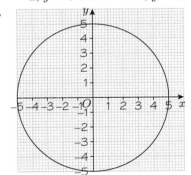

  b $x = 3.35, y = 3.7$ and $x = -0.95, y = -4.9$

Skills boost 2 Solving linear and non-linear simultaneous equations algebraically

Guided practice

a
$$2x + 3 = 2x^2 - 3x - 9$$
$$2x^2 - 5x - 12 = 0$$
$$(2x + 3)(\underline{\mathbf{x - 4}}) = 0$$

Either $(2x + 3) = 0$, giving $x = -\frac{3}{2}$

or $(x - 4) = 0$, giving $x = \underline{\mathbf{4}}$

When $x = -\frac{3}{2}, y = 2(-\frac{3}{2}) + 3 = 0$

When $x = \underline{\mathbf{4}}, y = 2(\mathbf{4}) + 3 = \underline{\mathbf{11}}$

So the solutions are $x = -\frac{3}{2}, y = 0$ and $x = \underline{\mathbf{4}}$, $y = \underline{\mathbf{11}}$

b $5x - y + 13 = 0$
$$y = 5x + 13$$
$$x^2 + (5x + 13)^2 = 13$$
$$x^2 + 25x^2 + 130x + 169 = 13$$
$$26x^2 + 130x + 156 = 0$$
$$x^2 + 5x + 6 = 0$$
$$(x + 2)(\underline{\mathbf{x + 3}}) = 0$$

Either $(x + 2) = 0$, giving $x = -2$

or $(\underline{\mathbf{x + 3}}) = 0$, giving $x = \underline{\mathbf{-3}}$

When $x = -2, y = 5(-2) + 13 = 3$

When $x = \underline{\mathbf{-3}}, y = \underline{\mathbf{-2}}$

So the solutions are $x = -2, y = 3$ and $x = \underline{\mathbf{-3}}$, $y = \underline{\mathbf{-2}}$

① **a** $x = 7, y = 10$ and $x = -2, y = 1$

  b $x = -3, y = -6$ and $x = 5, y = 10$

  c $x = -4, y = -6$ and $x = 1, y = -1$

  d $x = -0.69, y = 3.69$ and $x = 2.19, y = 0.81$

  e $x = \frac{2}{3}, y = 1\frac{1}{3}$ and $x = -1, y = 3$

  f $x = -\frac{3}{4}, y = 0$ and $x = 2, y = 11$

② **a** $x = -6, y = 8$ and $x = 5, y = -3$

  b $x = \frac{2}{3}, y = 0$ and $x = -6, y = 20$

  c $x = -2, y = 2$ and $x = -\frac{1}{5}, y = -\frac{8}{5}$

③ **a** $x = 0$ $y = \frac{1}{2}$ and $x = 14, y = 4$

  b $x = 58, y = -8$ and $x = -5, y = 2\frac{1}{2}$

  c $x = 7\frac{2}{3}, y = 1\frac{1}{3}$ and $x = -1\frac{1}{2}, y = -\frac{1}{2}$

④ **a** $x = 0.271, y = 1.65$ and $x = -2.77, y = 16.9$

  b $x = 2.54, y = 19.3$ and $x = 0.131, y = 0.0519$

  c $x = 1.16, y = -0.28$ and $x = 7.34, y = 1.78$

⑤ **a** $x = 4, y = 3$ and $x = -3, y = -4$

  b $x = -3, y = 1$ and $x = 1, y = -3$

  c $x = 6, y = 8$ and $x = -8, y = 6$

  d $x = 4, y = 5$ and $x = 5, y = 4$

  e $x = 3, y = -6$ and $x = 6, y = -3$

  f $x = 2, y = 6$ and $x = 6, y = -2$

⑥ **a** $x = 2, y = 3$ and $x = -2, y = -3$

  b $x = 12, y = 5$ and $x = -12, y = -5$

  c $x = 8, y = -1$ and $x = 1, y = 8$

  d $x = -5, y = 7$ and $x = 7, y = 5$

  e $x = 9, y = -1$ and $x = -9, y = 1$

  f $x = -6, y = 9$ and $x = 9, y = 6$

⑦ **a** $x = 2.04, y = 5.08$ and $x = -2.84, y = -4.68$

  b $x = 4.31, y = -5.61$ and $x = -1.90, y = 6.81$

  c $x = 2.87, y = -0.871$ and $x = -0.871, y = 2.87$

⑧ $x = -0.39, y = 0.23$ and $x = 1.72, y = 4.44$

⑨ $x = 10, y = 5$ and $x = 5, y = 10$

⑩ $x = -1.90, y = 6.81$ and $x = 4.31, y = -5.61$

Practise the methods

① **a** **i** $x^2 + 3x - 4 = 0$

  ii $x = -4, y = -8$ and $x = 1, y = 2$

  b **i** $x^2 + 2x - 8 = 0$

  ii $x = -4, y = -10$ and $x = 2, y = 8$

c i $2x^2 - x - 1 = 0$

   ii $x = -\frac{1}{2}, y = 3\frac{1}{2}$ and $x = 1, y = 5$

②a i $x^2 - x - 6 = 0$

   ii $x = -2, y = -3$ and $x = 3, y = 2$

 b i $x^2 + 7x - 30 = 0$

   ii $x = -10, y = -3$ and $x = 3, y = 10$

 c i $x^2 + 3x + 2 = 0$

   ii $x = -1, y = -2$ and $x = -2, y = 1$

③ $x = (-3, 3)\ y = -5$ and $x = 2\frac{1}{2}, y = 11\frac{1}{2}$

④ $x = -1.7, y = 4.7$ and $x = 4.7, y = -1.7$

Problem-solve!

① 22 metres

② $\sqrt{5}$

③ $(-3, 3)$ and $(7, 23)$

④ $x = 1.56, y = 3.68$ and $x = -0.96, y = -3.88$

Unit 5 Trigonometric graphs

①a 0.940   **b** 1.19   **c** 0.966   **d** 5.67

②a $\sqrt{3}$ cm

 b

sin 30°	sin 60°	cos 30°	cos 60°	tan 30°	tan 60°
$\frac{1}{2}$	$\frac{\sqrt{3}}{2}$	$\frac{\sqrt{3}}{2}$	$\frac{1}{2}$	$\frac{1}{\sqrt{3}}$	$\sqrt{3}$

③a $\sqrt{2}$

 b

sin 45°	cos 45°	tan 45°
$\frac{1}{\sqrt{2}}$	$\frac{1}{\sqrt{2}}$	1

Confidence check

①
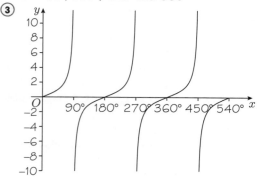

$x = 30°$ and $150°$

②

$x = 60°, 300°, 420°$ and $660°$

③

$x = 135°, 315°$ and $495°$

④ $x = 78.5°$ and $-78.5°$

Skills boost 1 Graph of the sine function

a $\frac{540}{360} = 1\frac{1}{2}$ cycles

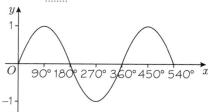

b $\sin x = \frac{1}{\sqrt{2}}$

   $x = 45°$

From the graph, there are three other values for which

$\sin x = \frac{1}{\sqrt{2}}$

$x = 180° - 45°$

 $= 135°$

$x = 45° + 360°$

 $= 405°$

$x = 540° - 45°$

 $= 495°$

When $\sin x = \frac{1}{\sqrt{2}}$, $x = 45°, 135°, 405°, 495°$

① C

②a 180° or 360°   **b** 300°

 c 210°   **d** 105°

③a
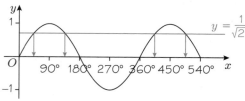

 b i −1   **ii** 0.5   **iii** −0.5

 c $x = 60°, 120°, 420°, 480°$

④a

 b $x = -330°, -210°, 30°, 150°$

⑤ A(−90°, −1), B(180°, 0), C(330°, −0.5), D(450°, 1)

Skills boost 2 Graph of the cosine function

a $\frac{720}{360} = 2$ cycles

b $\cos x = \dfrac{\sqrt{3}}{2}$

   $x = 30°$

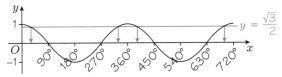

From the graph, there are three other values for which

$\cos x = \dfrac{\sqrt{3}}{2}$

$x = 360° - 30°$

   $= \underline{330°}$

$x = 30° + \underline{360°}$

   $= \underline{390°}$

$x = 720° - \underline{30°}$

   $= \underline{690°}$

When $\cos x = \dfrac{\sqrt{3}}{2}$, $x = 30°, 330°, 390°, 690°$

① B

② **a** 270°   **b** 240°   **c** 330°   **d** 285°

③ **a**

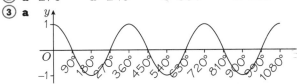

b **i** −1   **ii** −0.5   **iii** 0.5

c $x = 60°, 300°, 420°, 660°, 780°, 1020°$

④ **a**

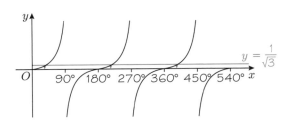

b $x = -315°, -45°, 45°, 315°$

⑤ A(−180°, −1), B(270°, 0), C(300°, $\frac{1}{2}$) and D(360°, ˆ

Skills boost 3 Graph of the tangent function

Guided practice

a $\dfrac{540}{180} = \underline{3}$ cycles

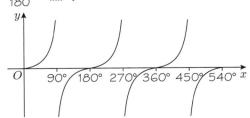

b $\tan x = \dfrac{1}{\sqrt{3}}$

   $x = 30°$

From the graph, there are two other values for which

$\tan x = \dfrac{1}{\sqrt{3}}$

$x = 30° + 180°$

   $= \underline{210°}$

$x = 30° + \underline{360°}$

   $= \underline{390°}$

When $\tan x = \dfrac{1}{\sqrt{3}}$, $x = 30°, 210°, 390°$

① B

② **a** 180° or 360°

   b 225°

   c 300°

   d 285°

③ **a**

b **i** 0   **ii** −1   **iii** 1

c $x = 45°, 225°, 405°, 585°$

④ **a**

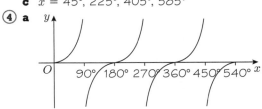

b $x = 45°, 225°, 405°$

⑤ **a**

b $x = -300°, -120°, 60°, 240°$

Skills boost 4 Solving trigonometric equations

Guided practice

$4 \sin x = 3$

$\sin x = \dfrac{3}{4}$

   $x = \sin^{-1}\left(\dfrac{3}{4}\right)$

   $= 48.6°$ (to 3 s.f.)

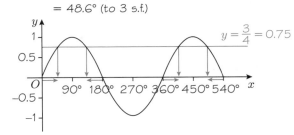

From the graph, the other values of x are:

$180° - 48.6° = \textbf{131.4°}$

$360° + 48.6° = \textbf{408.6°}$

$\underline{540°} - 48.6° = \textbf{491.4°}$

$x = 48.6°, 131.4°, 408.6°, 491.4°$

① $x = 23.6°, 156.4°, 383.6°, 516.4°$

② $x = -308.7°, -51.3°, 51.3°, 308.7°$

③ $x = 55.0°, 235.0°, 415.0°$

④ $x = 67.4°, 247.4°, 427.4°, 607.4°$

5 a

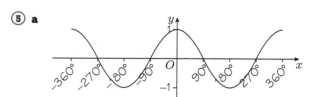

b $x = -326.4°, -33.6°, 33.6°$ and $326.4°$

Practise the methods

1

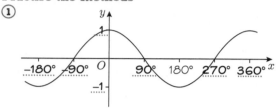

2 A(90°, 1), B(270°, −1), C(360°, 0) and D(−270°, 1)

3 $x = 75.1°, 255.1°, 435.1°, 615.1°$

Problem-solve!

Examples are:

1
  a −270°, 90°, 450°, 810°
  b −390°, −30°, 330°, 690°
  c −540°, −180°, 180°, 540°
  d −480°, −120°, 240°, 600°
  e −225°, −45°, 135°, 315°
  f −300°, −120°, 60°, 240°

2

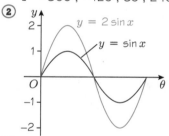

3 $x = 99.6°, 260.6°$
4 $x = -155.4°, -24.6°, 204.6°, 335.4°$
5 $x = -30°, 150°$
6 $x = 230°, 290°$

Unit 6 Functions

A01 Fluency check

1
  a $x = \dfrac{y - 7}{3}$
  b $x = \dfrac{3y + 1}{2}$
  c $x = \dfrac{y - 7}{4}$

2
  a $y = 15t$
  b $y = 9t^2$
  c $y = t^2 + t$

Confidence check

1
  a $6x + 2$
  b $8x^2 - 3$
  c $2x^2 + 3x - 2$

2 $\dfrac{x}{3} - 2$

3 $4x^2 - 20x + 28$

Skills boost 1 Using function notation

Guided practice

a $3f(x) = \underline{3(4x - 3)}$
        $= 12x - 9$
b $g(-2x) = 3\underline{(-2x)}^2 + 5$
           $= 12x^2 + 5$
c $f(x) + g(x) = 4x - 3 + \underline{3x^2 + 5}$
              $= 3x^2 + 4x + 2$

1
  a 16
  b −2
  c 3
  d 3
  e 54
  f −2
  g 34
  h 19

2
  a $4x + 10$
  b $4x + 2$
  c $8x + 14$
  d $16x + 28$
  e $12x + 7$
  f $20x + 7$

3
  a $5x + 3$
  b $5x - 4$
  c $15x - 3$
  d $10x - 1$
  e $5x + 9$
  f $-5x - 1$

4
  a $x^2 + 4$
  b $x^2 - 1$
  c $2x^2 + 6$
  d $4x^2 + 3$
  e $x^2 + 3$
  f $x^2 + 2x + 4$

5
  a $2x^2 - 2$
  b $2x^2 - 6$
  c $6x^2 - 15$
  d $18x^2 - 5$
  e $2x^2 + 8x + 3$
  f $2x^2 - 4x - 3$

6 $2x^2 + 13x$

7
  a $7x + 5$
  b $4x - 3$
  c $3x + 6$
  d $x^2 + x - 4$
  e $2x^2 + 3x + 1$

8
  a $2x - 2$
  b $-2x + 2$

9
  a $-3x + 4$
  b $3x - 4$

10
  a $x^2 - x + 7$
  b $-4x^2 - x + 4$

Skills boost 2 Inverse functions

Guided practice

a $y = 2x + 5$
      $2x = y - \underline{5}$
      $x = \dfrac{y - 5}{2}$

The inverse function of $x \to 2x + 5$ is $x \to \dfrac{x - 5}{2}$

b $y = \dfrac{2x}{3} - 1$
        $y + 1 = \dfrac{2x}{3}$
        $\underline{3}(y + 1) = 2x$
        $x = \dfrac{3(y + 1)}{2}$
        $f^{-1}(x) = \dfrac{3(x + 1)}{2}$

1
  a $x \to \dfrac{x - 3}{4}$
  b $x \to \dfrac{x}{3} + 5$
  c $x \to 4(x - 1)$
  d $x \to 5x + 2$

2
  a $f^{-1}(x) = \dfrac{x}{2} - 3$
  b $g^{-1}(x) = \dfrac{x}{3} - 1$
  c $h^{-1}(x) = \dfrac{3(x + 5)}{4}$

3 $f^{-1}(x) = \dfrac{x + 1}{3}$

Skills boost 3 Composite functions

Guided practice

a $g(6) = \underline{6}^2 + 5$
       $= \underline{41}$
  $fg(6) = 3 - 2(\underline{41})$
        $= -79$
b $fg(x) = 3 - 2(\underline{x^2 + 5})$
        $= 3 - \underline{2x^2 - 10}$
        $= -2x^2 - 7$
c $gf(x) = \underline{(3 - 2x)}^2 + 5$
       $= \underline{9 - 12x + 4x^2} + 5$
       $= 4x^2 - 12x + 14$

1
  a 5
  b −7
  c 3

2
  a $21 - 10x$
  b $-17 - 10x$
  c $4x + 15$

3 a $3x^2 + 7$ **b** $9x^2 + 6x + 3$ **c** $9x + 7$

4 a $-4x + 2$ **b** $23 - 4x$
  c $14 - 2x^2$ **d** $32x^2 + 48x + 9$
  e $2x^2 - 20x + 41$ **f** $8x^2 - 33$

5 a $11 - 4x$ **b** $4 - 4x$
  c $7 - 4x^2$ **d** $-x^2 + 6x - 7$
  e $-16x^2 + 8x + 1$ **f** $x^2 + 1$

6 $f(x) = 2(x - 3) = 2x - 6$
  $ff(x) = 2(2x - 6 - 3) = 2(2x - 9) = 4x - 18$

Practise the methods

1 $f^{-1}(x) = \dfrac{7(x + 4)}{5}$

2 a $f^{-1}(x) = \dfrac{x + 7}{3}$ **b** $f^{-1}(x) = \dfrac{x}{4} - 9$
  c $f^{-1}(x) = 3x + 8$ **d** $f^{-1}(x) = 5(x - 2)$

3 $f^{-1}(x) = \dfrac{x - 9}{5}$

4 a $8 - 5x$ **b** $12 - 10x$
  c $6 - 15x$ **d** $-4 - 5x$
  e $2x^2 - 4x - 3$ **f** $2x^2 - 5x + 1$

5 $f(x + 2) = 3(x + 2)^2 - (x + 2) + 4$
    $= 3(x^2 + 4x + 4) - x - 2 + 4$
    $= 3x^2 + 12x + 12 - x + 2$
    $= 3x^2 + 11x + 14$

6 $f(x) = 3x + 3$ and $g(x) = 3x - 6$
  $fg(x) = 3(3x - 6) + 3 = 9x - 18 + 3 = 9x - 15$

Problem-solve!

1 $a = 2\dfrac{1}{2}$

2 a $f(1) = 1 - 3 \times 1 = -2$
    $g(-2) = 3 \times (-2) - 1 = -7$
  b $f^{-1}(x) = \dfrac{1}{3} - \dfrac{x}{3}$ and $g^{-1}(x) = \dfrac{x}{3} + \dfrac{1}{3}$
    $f^{-1}(x) + g^{-1}(x) = \dfrac{1}{3} - \dfrac{x}{3} + \dfrac{x}{3} + \dfrac{1}{3} = \dfrac{2}{3}$

3 a $f^{-1}(x) = \dfrac{x + 1}{2}$ **b** $k = 5$

4 a i $x^2 + 3$ **ii** $x^2 - 10x + 33$
  b $x = 3$

5 a i x **ii** x
  b $f(x)$ and $g(x)$ are inverse functions because
    $fg(x) = gf(x) = x$

6 $f(x)$ and $g(x)$ are not inverse functions because
  $fg(x) = x - \dfrac{25}{6}$ and $gf(x) = x - 25$, therefore
  $fg(x) \neq gf(x) \neq x$

Unit 7 Transformations of graphs

1 a 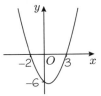   **b**

  c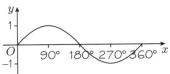

d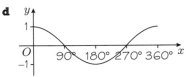

2 a $(3, 4)$ **b** $(4, -2)$

Confidence check

1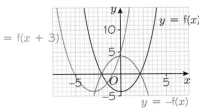

2 $x^2 + 6x + 5$

Skills boost 1 Transforming graphs

Guided practice

a i, ii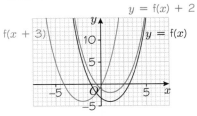

b i, ii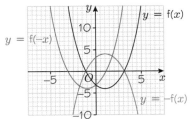

c i, ii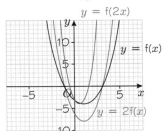

1 a, b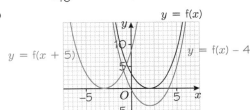

2 a, b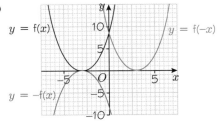

③ a, b

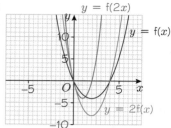

④ a

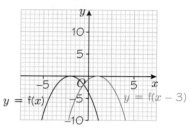

b

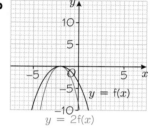

⑤ a, b

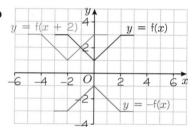

⑥ a

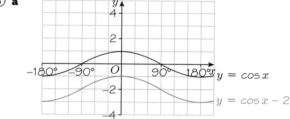

b

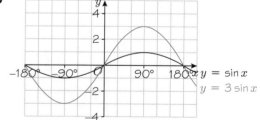

Skills boost 2 Interpreting transformations of graphs

Guided practice

The graph is the same but moved to the left and up, so the transformation is a **translation**.

Graph A is a translation of f(x) by **3** units left
and **2** units up.

Graph A = f(x + 3) + 2
= ($\underline{x + 3}$)^2 + **2**
= $x^2 + 6x + 11$

**① ** $y = -(x + 1)^3 + 3$

**② ** $y = x^2 + 8x + 16$ or $y = (x + 4)^2$

Practise the methods

① a i

x	−2	−1	0	1	2	3
f(x)	−7	−5	−3	−1	1	3

ii

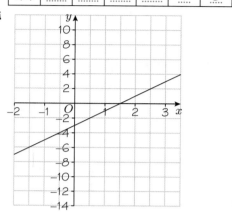

b i

x	−2	−1	0	1	2	3
2f(x)	−14	−10	−6	−2	2	6

ii The values of 2f(x) are double the values of f(x).

iii

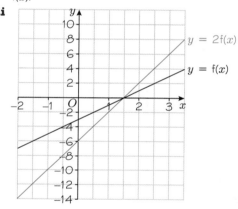

iv $y = 2$f(x) is a vertical stretch of $y =$ f(x) of scale factor 2.

c i

x	−2	−1	0	1	2	3
f(2x)	−11	−7	−3	1	5	9

ii

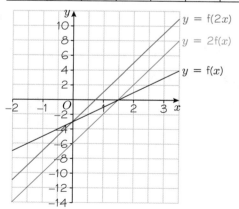

iii $y =$ f(2x) is a horizontal stretch of $y =$ f(x) of scale factor $\frac{1}{2}$.

② a $y =$ f(x) + 7 is a translation of $y =$ f(x) by $\begin{pmatrix} 0 \\ 7 \end{pmatrix}$.

b $y =$ f(x − 3) is a translation of $y =$ f(x) by $\begin{pmatrix} 3 \\ 0 \end{pmatrix}$.

c $y = f(-x)$ is a reflection of $y = f(x)$ in the y-axis

d $y = f(3x)$ is a horizontal stretch of $y = f(x)$ scale factor $\frac{1}{3}$.

e $y = 5f(x)$ is a vertical stretch scale factor 5 of $y = f(x)$.

③ a, b

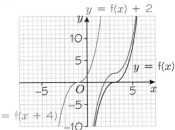

④ a, b

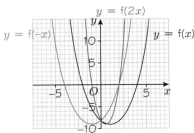

⑤ $y = 3 - x^2$

Problem-solve!

① $y = -g(x)$ or $y = -(x - 1)(x - 5)$

② $(9, -2)$

③

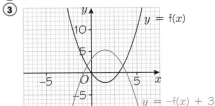

④

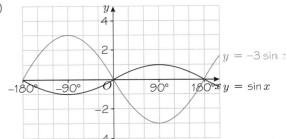

⑤ $a = -2$, $b = 2$ and $c = -1$

Unit 8 Pre-calculus

A01 Fluency check

① 24 cm²

② Gradient of line A $= -\frac{1}{2}$

Gradient of line B $= -1$

Gradient of line C $= 3$

Gradient of line D $= 2$

Confidence check

① Speed

② 28.5 units²

Skills boost 1 Gradient

Guided practice

a Gradient $= \dfrac{\text{change in speed}}{\text{change in } t}$

Students' values will vary depending on the accuracy of their gradient, an example may be

$$\frac{5.8 - 0.2}{6 - 2} = 1.4$$

b 4 seconds after the start of the race, Lexi was accelerating at 1.4 m/s².

① **a** Gradient $= 6.5$

b 3 seconds after setting off, the cyclist was cycling at a speed of 6.5 m/s.

Skills boost 2 Area under a curve

Guided practice

Area of trapezium A $= \dfrac{1}{2} \times 2(21 + \underline{25})$

$\qquad\qquad\qquad\qquad = \underline{46}$

Area of trapezium B $= \dfrac{1}{2} \times \underline{2}\,(\underline{25} + \underline{27})$

$\qquad\qquad\qquad\qquad = \underline{52}$

Total area $= \underline{46} + \underline{52}$

$\qquad\qquad = 98$

An estimate for the distance the car travelled between 4 and 8 seconds is 98 metres.

① Estimates between 450 and 465 m

② **a**

b 21 units²

③ 700 m

Practise the methods

① **a** 4   **b** −2   **c** −6

② **a** 14 m/s²   **b** 90 m

Problem-solve!

① −1.1 at (3, 2.65) or 1.1 at (3, −2.65)

② **a** Abi is leading the race throughout. Abi started the race fast and then slowed down slightly towards the end. She finished in 44 minutes. Bella started slowly, but then sped up, but not enough to catch Abi. Bella finished in 47.5 minutes.

b Estimating from the graph they are both going at the same speed. around 32 km/h.

Difference $= 0$ km/h

③ 70 units²